T0260033

THERMODYNAMIC DEGRADATION SCIENCE

Wiley Series in Quality & Reliability Engineering

Dr Andre Kleyner
Series Editor

The Wiley series in Quality & Reliability Engineering aims to provide a solid educational foundation for both practitioners and researchers in Q&R field and to expand the reader's knowledge base to include the latest developments in this field. The series will provide a lasting and positive contribution to the teaching and practice of engineering.

The series coverage will contain, but is not exclusive to,

- statistical methods;
- physics of failure;
- reliability modeling;
- functional safety;
- six-sigma methods;
- lead-free electronics;
- warranty analysis/management; and
- risk and safety analysis.

Wiley Series in Quality & Reliability Engineering

Next Generation HALT and HASS: Robust Design of Electronics and Systems
by Kirk A. Gray, John J. Paschkewitz
May 2016

Reliability and Risk Models: Setting Reliability Requirements, 2nd Edition
by Michael Todinov
September 2015

Applied Reliability Engineering and Risk Analysis: Probabilistic Models and Statistical Inference
by Ilia B. Frenkel, Alex Karagrigoriou, Anatoly Lisnianski, Andre V. Kleyner
September 2013

Design for Reliability
by Dev G. Raheja (Editor), Louis J. Gullo (Editor)
July 2012

Effective FMEAs: Achieving Safe, Reliable, and Economical Products and Processes using Failure Mode and Effects Analysis
by Carl Carlson
April 2012

Failure Analysis: A Practical Guide for Manufacturers of Electronic Components and Systems
by Marius Bazu, Titu Bajenescu
April 2011

Reliability Technology: Principles and Practice of Failure Prevention in Electronic Systems
by Norman Pascoe
April 2011

Improving Product Reliability: Strategies and Implementation
by Mark A. Levin, Ted T. Kalal
March 2003

Test Engineering: A Concise Guide to Cost-effective Design, Development and Manufacture
by Patrick O'Connor
April 2001

Integrated Circuit Failure Analysis: A Guide to Preparation Techniques
by Friedrich Beck
January 1998

Measurement and Calibration Requirements for Quality Assurance to ISO 9000
by Alan S. Morris
October 1997

Electronic Component Reliability: Fundamentals, Modelling, Evaluation, and Assurance
by Finn Jensen
November 1995

THERMODYNAMIC DEGRADATION SCIENCE

PHYSICS OF FAILURE, ACCELERATED TESTING, FATIGUE, AND RELIABILITY APPLICATIONS

Alec Feinberg, Ph.D.
DfRSoftware Company, Raleigh, NC, USA

WILEY

Registered Office
John Wiley & Sons, Ltd, The Atrium, Southern Gate, Chichester, West Sussex, PO19 8SQ, United Kingdom

For details of our global editorial offices, for customer services and for information about how to apply for permission to reuse the copyright material in this book please see our website at www.wiley.com.

Library of Congress Cataloging-in-Publication Data

Names: Feinberg, Alec, author
Title: Thermodynamic degradation science : physics of failure, accelerated testing, fatigue and reliability applications / Alec Feinberg, Ph.D.
Description: Hoboken, NJ : John Wiley & Sons, Inc., [2016] | Series: Wiley series in quality and reliability engineering | Includes bibliographical references and index.
Identifiers: LCCN 2016017320 (print) | LCCN 2016031239 (ebook) | ISBN 9781119276227 (cloth) | ISBN 9781119276241 (pdf) | ISBN 9781119276272 (epub)
Subjects: LCSH: Heat-engines–Thermodynamics. | Metals–Fatigue. |Metals–Testing. | Thermodynamic equilibrium.
Classification: LCC TJ265 .F45 2016 (print) | LCC TJ265 (ebook) | DDC 620.1/61–dc23
LC record available at https://lccn.loc.gov/2016017320

A catalogue record for this book is available from the British Library.

Cover image: Gettyimages/AlexSava

Set in 10/12pt Times by SPi Global, Pondicherry, India
Printed and bound in Malaysia by Vivar Printing Sdn Bhd

1 2016

To Linda

Failure is Not an Option

In many situations, failure is not an option. It can take immense planning to prevent failure. Thermodynamic degradation science offers new tools and measurement methods that can help.

Second Law of Thermodynamics in Terms of Aging

The spontaneous irreversible degradation processes that take place in a system interacting with its environment will do so in order to go towards thermodynamic equilibrium with its environment.

Entropy Damage

The entropy generated associated with system degradation is "entropy damage."

$$\Delta S_{\text{system}} = \Delta S_{\text{damage}} + \Delta S_{\text{non-damage}}, \quad \Delta S_{\text{damage}} \geq 0$$

$$W_{\text{actual}} = W_{\text{rev}} - W_{\text{irr}}$$

$$\text{Cum damage} = \frac{\sum_n \oint Y_n dX_n}{W_{\text{failure}}}$$

The Four Main Aging Categories

- Forced processes;
- Activation;
- Diffusion; and
- Combinations of these, yielding complex aging.

Contents

List of Figures xiii
List of Tables xvi
About the Author xvii
Preface xviii

1 Equilibrium Thermodynamic Degradation Science 1
1.1 Introduction to a New Science 1
1.2 Categorizing Physics of Failure Mechanisms 2
1.3 Entropy Damage Concept 3
1.3.1 The System (Device) and its Environment 4
1.3.2 Irreversible Thermodynamic Processes Cause Damage 5
1.4 Thermodynamic Work 6
1.5 Thermodynamic State Variables and their Characteristics 7
1.6 Thermodynamic Second Law in Terms of System Entropy Damage 9
1.6.1 Thermodynamic Entropy Damage Axiom 11
1.6.2 Entropy and Free Energy 13
1.7 Work, Resistance, Generated Entropy, and the Second Law 14
1.8 Thermodynamic Catastrophic and Parametric Failure 16
1.8.1 Equilibrium and Non-Equilibrium Aging States in Terms of the Free Energy or Entropy Change 16
1.9 Repair Entropy 17
1.9.1 Example 1.1: Repair Entropy: Relating Non-Damage Entropy Flow to Entropy Damage 17
Summary 18
References 22

2 Applications of Equilibrium Thermodynamic Degradation to Complex and Simple Systems: Entropy Damage, Vibration, Temperature, Noise Analysis, and Thermodynamic Potentials 23
2.1 Cumulative Entropy Damage Approach in Physics of Failure 23
2.1.1 Example 2.1: Miner's Rule Derivation 25
2.1.2 Example 2.2: Miner's Rule Example 26
2.1.3 Non-Cyclic Applications of Cumulative Damage 27
2.2 Measuring Entropy Damage Processes 27
2.3 Intermediate Thermodynamic Aging States and Sampling 29
2.4 Measures for System-Level Entropy Damage 29
2.4.1 Measuring System Entropy Damage with Temperature 29
2.4.2 Example 2.3: Resistor Aging 30
2.4.3 Example 2.4: Complex Resistor Bank 31
2.4.4 System Entropy Damage with Temperature Observations 32
2.4.5 Example 2.5: Temperature Aging of an Operating System 32
2.4.6 Comment on High-Temperature Aging for Operating and Non-Operating Systems 32
2.5 Measuring Randomness due to System Entropy Damage with Mesoscopic Noise Analysis in an Operating System 33
2.5.1 Example 2.6: Gaussian Noise Vibration Damage 35
2.5.2 Example 2.7: System Vibration Damage Observed with Noise Analysis 36
2.6 How System Entropy Damage Leads to Random Processes 37
2.6.1 Stationary versus Non-Stationary Entropy Process 40
2.7 Example 2.8: Human Heart Rate Noise Degradation 41
2.8 Entropy Damage Noise Assessment Using Autocorrelation and the Power Spectral Density 42
2.8.1 Noise Measurements Rules of Thumb for the PSD and R 43
2.8.2 Literature Review of Traditional Noise Measurement 44
2.8.3 Literature Review for Resistor Noise 48
2.9 Noise Detection Measurement System 48
2.9.1 System Noise Temperature 49
2.9.2 Environmental Noise Due to Pollution 50
2.9.3 Measuring System Entropy Damage using Failure Rate 50
2.10 Entropy Maximize Principle: Combined First and Second Law 51
2.10.1 Example 2.9: Thermal Equilibrium 52
2.10.2 Example 2.10: Equilibrium with Charge Exchange 53
2.10.3 Example 2.11: Diffusion Equilibrium 55
2.10.4 Example 2.12: Available Work 55
2.11 Thermodynamic Potentials and Energy States 57
2.11.1 The Helmholtz Free Energy 58
2.11.2 The Enthalpy Energy State 60
2.11.3 The Gibbs Free Energy 60
2.11.4 Summary of Common Thermodynamic State Energies 62
2.11.5 Example 2.13: Work, Entropy Damage, and Free Energy Change 62
2.11.6 Example 2.14: System in Contact with a Reservoir 65

Summary 68
References 76

3 NE Thermodynamic Degradation Science Assessment Using the Work Concept 77
3.1 Equilibrium versus Non-Equilibrium Aging Approach 77
3.1.1 Conjugate Work and Free Energy Approach to Understanding Non-Equilibrium Thermodynamic Degradation 78
3.2 Application to Cyclic Work and Cumulative Damage 79
3.3 Cyclic Work Process, Heat Engines, and the Carnot Cycle 81
3.4 Example 3.1: Cyclic Engine Damage Quantified Using Efficiency 84
3.5 The Thermodynamic Damage Ratio Method for Tracking Degradation 86
3.6 Acceleration Factors from the Damage Ratio Principle 87
Summary 89
References 92

4 Applications of NE Thermodynamic Degradation Science to Mechanical Systems: Accelerated Test and CAST Equations, Miner's Rule, and FDS 93
4.1 Thermodynamic Work Approach to Physics of Failure Problems 93
4.2 Example 4.1: Miner's Rule 93
4.2.1 Acceleration Factor Modification of Miner's Damage Rule 95
4.3 Assessing Thermodynamic Damage in Mechanical Systems 96
4.3.1 Example 4.2: Creep Cumulative Damage and Acceleration Factors 96
4.3.2 Example 4.3: Wear Cumulative Damage and Acceleration Factors 99
4.3.3 Example 4.4: Thermal Cycle Fatigue and Acceleration Factors 101
4.3.4 Example 4.5: Mechanical Cycle Vibration Fatigue and Acceleration Factors 102
4.3.5 Example 4.6: Cycles to Failure under a Resonance Condition: Q Effect 105
4.4 Cumulative Damage Accelerated Stress Test Goal: Environmental Profiling and Cumulative Accelerated Stress Test (CAST) Equations 107
4.5 Fatigue Damage Spectrum Analysis for Vibration Accelerated Testing 108
4.5.1 Fatigue Damage Spectrum for Sine Vibration Accelerated Testing 109
4.5.2 Fatigue Damage Spectrum for Random Vibration Accelerated Testing 110
Summary 111
References 117

5 Corrosion Applications in NE Thermodynamic Degradation 118
5.1 Corrosion Damage in Electrochemistry 118
5.1.1 Example 5.1: Miner's Rule for Secondary Batteries 119
5.2 Example 5.2: Chemical Corrosion Processes 121
5.2.1 Example 5.3: Numerical Example of Linear Corrosion 123
5.2.2 Example 5.4: Corrosion Rate Comparison of Different Metals 124
5.2.3 Thermal Arrhenius Activation and Peukert's Law 124
5.3 Corrosion Current in Primary Batteries 126
5.3.1 Equilibrium Thermodynamic Condition: Nernst Equation 127

5.4 Corrosion Rate in Microelectronics 128
5.4.1 Corrosion and Chemical Rate Processes Due to Temperature 129
Summary 130
References 133

6 Thermal Activation Free Energy Approach 134
6.1 Free Energy Roller Coaster 134
6.2 Thermally Activated Time-Dependent (TAT) Degradation Model 135
6.2.1 Arrhenius Aging Due to Small Parametric Change 136
6.3 Free Energy Use in Parametric Degradation and the Partition Function 138
6.4 Parametric Aging at End of Life Due to the Arrhenius Mechanism: Large Parametric Change 140
Summary 141
References 143

7 TAT Model Applications: Wear, Creep, and Transistor Aging 144
7.1 Solving Physics of Failure Problems with the TAT Model 144
7.2 Example 7.1: Activation Wear 144
7.3 Example 7.2: Activation Creep Model 146
7.4 Transistor Aging 148
7.4.1 Bipolar Transistor Beta Aging Mechanism 148
7.4.2 Capacitor Leakage Model for Base Leakage Current 149
7.4.3 Thermally Activated Time-Dependent Model for Transistors and Dielectric Leakage 150
7.4.4 Field-Effect Transistor Parameter Degradation 152
Summary 154
References 156

8 Diffusion 157
8.1 The Diffusion Process 157
8.2 Example 8.1: Describing Diffusion Using Equilibrium Thermodynamics 157
8.3 Describing Diffusion Using Probability 159
8.4 Diffusion Acceleration Factor with and without Temperature Dependence 161
8.5 Diffusion Entropy Damage 161
8.5.1 Example 8.2: Package Moisture Diffusion 162
8.6 General Form of the Diffusion Equation 163
Summary 164
Reference 166

9 How Aging Laws Influence Parametric and Catastrophic Reliability Distributions 167
9.1 Physics of Failure Influence on Reliability Distributions 167
9.2 Log Time Aging (or Power Aging Laws) and the Lognormal Distribution 168
9.3 Aging Power Laws and the Weibull Distribution: Influence on Beta 171

9.4 Stress and Life Distributions 175
9.4.1 Example 9.1: Cumulative Distribution Function as a Function of Stress 176
9.5 Time- (or Stress-) Dependent Standard Deviation 177
Summary 178
References 180

10 The Theory of Organization: Final Thoughts **181**

Special Topics A: Key Reliability Statistics 183
A.1 Introduction 183
A.1.1 Reliability and Accelerated Testing Software to Aid the Reader 183
A.2 The Key Reliability Functions 184
A.3 More Information on the Failure Rate 186
A.4 The Bathtub Curve and Reliability Distributions 187
A.4.1 Exponential Distribution 188
A.4.2 Weibull Distribution 190
A.4.3 Normal (Gaussian) Distribution 191
A.4.4 The Lognormal Reliability Function 194
A.5 Confidence Interval for Normal Parametric Analysis 195
A.5.1 Example A.4: Power Amplifier Confidence Interval 196
A.6 Central Limit Theorem and Cpk Analysis 197
A.6.1 Cpk Analysis 197
A.6.2 Example A.5: Cpk and Yield for the Power Amplifiers 197
A.7 Catastrophic Analysis 199
A.7.1 Censored Data 199
A.7.2 Example A.6: Weibull and Lognormal Analysis of Semiconductors 199
A.7.3 Example A.7: Mixed Modal Analysis Inflection Point Method 201
A.8 Reliability Objectives and Confidence Testing 203
A.8.1 Chi-Squared Confidence Test Planning for Few Failures: The Exponential Case 204
A.8.2 Example A.8: Chi-Squared Accelerated Test Plan 205
A.9 Comprehensive Accelerated Test Planning 205
References 206

Special Topics B: Applications to Accelerated Testing 207
B.1 Introduction 207
B.1.1 Reliability and Accelerated Testing Software to Aid the Reader 208
B.1.2 Using the Arrhenius Acceleration Model for Temperature 209
B.1.3 Example B.2: Estimating the Activation Energy 211
B.1.4 Example B.3: Estimating Mean Time to Failure from Life Test 212
B.2 Power Law Acceleration Factors 212
B.2.1 Example B.4: Generalized Power Law Acceleration Factors 214
B.3 Temperature–Humidity Life Test Model 214
B.3.1 Temperature–Humidity Bias and Local Relative Humidity 215

B.4 Temperature Cycle Testing 216
B.4.1 Example B.6: Using the Temperature Cycle Model 217
B.5 Vibration Acceleration 217
B.5.1 Example B.7: Accelerated Testing Using Sine and Random Vibration 220
B.6 Multiple-Stress Accelerated Test Plans for Demonstrating Reliability 220
B.6.1 Example B.8: Designing Multi-Accelerated Tests Plans: Failure-Free 221
B.7 Cumulative Accelerated Stress Test (CAST) Goals and Equations Usage in Environmental Profiling 222
B.7.1 Example B.9: Cumulative Accelerated Stress Test (CAST) Goals and Equation in Environmental Profiling 222
References 223

Special Topics C: Negative Entropy and the Perfect Human Engine 224
C.1 Spontaneous Negative Entropy: Growth and Repair 224
C.2 The Perfect Human Engine: How to Live Longer 225
C.2.1 Differences and Similarities of the Human Engine to Other Systems 226
C.2.2 Knowledge of Cyclic Work to Improve Our Chances of a Longer Life 226
C.2.3 Example C.1: Exercise and the Human Heart Life Cycle 228
C.3 Growth and Self-Repair Part of the Human Engine 229
C.3.1 Example C.2: Work for Human Repair 230
C.4 Act of Spontaneous Negative Entropy 231
C.4.1 Repair Aging Rate: An RC Electrical Model 232
References 233

Overview of New Terms, Equations, and Concepts **234**

Index **236**

List of Figures

Figure 1.1	Conceptualized aging rates for physics-of-failure mechanisms	3
Figure 1.2	First law energy flow to system: (a) heat-in, work-out; and (b) heat-in and work-in	7
Figure 1.3	Fatigue *S–N* curve of cycles to failure versus stress, illustrating a fatigue limit in steel and no apparent limit in aluminum	10
Figure 1.4	Elastic stress limit and yielding point 1	12
Figure 2.1	The entropy change of an isolated system is the sum of the entropy changes of its components, and is never less than zero	24
Figure 2.2	Cell fatigue dislocations and cumulating entropy	25
Figure 2.3	Gaussian white noise	36
Figure 2.4	Noise limit heart rate variability measurements of young, elderly, and CHF patients [10]	41
Figure 2.5	Noise limit heart rate variability chaos measurements of young and CHF patients [10]	42
Figure 2.6	Graphical representation of the autocorrelation function	43
Figure 2.7	(a) Sine waves at 10 and 15 Hz with some randomness in frequency; and (b) Fourier transform spectrum. In (b) we cannot transform back without knowledge of which sine tone occurred first	44
Figure 2.8	(a) White noise time series; (b) normalized autocorrelation function of white noise; and (c) PSD spectrum of white noise	45
Figure 2.9	(a) Flicker (pink) $1/f$ noise; (b) normalized autocorrelation function of $1/f$ noise; and (c) PSD spectrum of $1/f$ noise	45
Figure 2.10	(a) Brown $1/f^2$ noise; (b) normalized autocorrelation function of $1/f^2$ noise; and (c) PSD spectrum of $1/f^2$ noise	46
Figure 2.11	Some key types of white, pink, and brown noise that might be observed from a system	47
Figure 2.12	$1/f$ noise simulations for resistor noise. Note the lower noise for larger resistors (power of 2) and higher noise for smaller resistors (power of 1.5)	48

Figure 2.13 Autocorrelation noise measurement detection system 49
Figure 2.14 Insulating cylinder divided into two sections by a frictionless piston 52
Figure 2.15 System (capacitor) and environment (battery) circuit 54
Figure 2.16 The system expands against the atmosphere 55
Figure 2.17 Mechanical work done on a system 63
Figure 2.18 Loss of available work due to increase in entropy damage 64
Figure 2.19 A simple system in contact with a heat reservoir 66
Figure 2.20 A system's free energy decrease over time and the corresponding total entropy increase 66
Figure 3.1 Conceptual view of cyclic cumulative damage 79
Figure 3.2 Cyclic work plane 80
Figure 3.3 Carnot cycle in P, V plane 81
Figure 3.4 Cyclic engine damage Area 1 > Area 2 85
Figure 4.1 Creep strain over time for different stresses where $\sigma_4 > \sigma_3 > \sigma_2 > \sigma_1$ 97
Figure 4.2 Example of creep of a wire due to a stress weight 97
Figure 4.3 Wear occurring to a sliding block having weight P_W 99
Figure 4.4 Graphical example of a sine test resonance 105
Figure 5.1 Lead acid and alkaline MnO_2 batteries fitted data 120
Figure 5.2 A simple corrosion cell with iron corrosion 124
Figure 5.3 Uniform electrochemical corrosion depicted on the surface of a metal 127
Figure 6.1 Arrhenius activation free energy path having a relative minimum as a function of generalized parameter a 135
Figure 6.2 Examples of $\ln(1 + B \text{ time})$ aging law, with upper graph similar to primary and secondary creep stages and the lower graph similar to primary battery voltage loss 137
Figure 6.3 Log time compared to power law aging models 138
Figure 6.4 (a) Continuous function with numerous energy states. (b) Relative minimum energy states having different degradation mechanisms 139
Figure 6.5 Aging with critical values t_c prior to catastrophic failure 141
Figure 7.1 Types of wear dependence on sliding distance (time) 145
Figure 7.2 Capacitor leakage model 149
Figure 7.3 Beta degradation on life test data 151
Figure 7.4 Life test data of gate-source MESFET leakage current over time fitted to the $\ln(1 + Bt)$ aging model. Junction rise was about 30°C 152
Figure 8.1 System with n particles and n_{env} environment particles 158
Figure 8.2 Diffusion concept 159
Figure 9.1 Reliability bathtub curve model 168
Figure 9.2 Power law fit to the wear-out portion of the bathtub curve 168
Figure 9.3 Log time aging with parametric threshold t_f 169
Figure 9.4 PDF failure portion that drifted past the parametric threshold 170
Figure 9.5 Creep curve with all three stages 172
Figure 9.6 Creep rate power law model for each creep stage, similar to the bathtub curve in Figure 9.1 173
Figure 9.7 Creep strain over time for different stresses where $\sigma_4 > \sigma_3 > \sigma_2 > \sigma_1$ 175
Figure 9.8 Crystal frequency drift showing time-dependent standard deviation 177

Figure A.1 Reliability bathtub curve model 187
Figure A.2 Demonstrating the power law on the wear-out shape 190
Figure A.3 Modeling the bathtub curve with the Weibull power law 191
Figure A.4 Weibull hazard (failure) rate for different values of β [1] 192
Figure A.5 Weibull shapes of PDF and CDF with $\beta = 2$ [1] 192
Figure A.6 Weibull shapes of PDF and CDF with $\beta = 0.5$ [1] 192
Figure A.7 Normal distribution shapes of PDF and CDF; $\mu = 5$, $\sigma = 1$ [1] 193
Figure A.8 Lognormal hazard (failure) rate for different σ values [1] 195
Figure A.9 Lognormal CDF and PDF for different σ values [1] 196
Figure A.10 Cpk analysis 198
Figure A.11 Life test: (a) Weibull analysis compared to (b) lognormal analysis test at 200°C [1] 200
Figure A.12 Field data (Table A.4) displaying inflection point as sub and main populations [1] 202
Figure A.13 Separating out the lower and upper distributions by the inflection point method [1] 203
Figure B.1 Main accelerated stresses and associated common failure issues 208
Figure B.2 Common accelerated qualification test plan used in industry [1] 209
Figure B.3 Arrhenius plot of data given in Table B.1 211
Figure B.4 MTTF stress plot of data given in Table B.2 214
Figure B.5 Sine vibration amplitude over time example 218
Figure B.6 Random vibration amplitude time series example 218
Figure B.7 PSD of the random vibration time series in Figure B.6 219
Figure C.1 *S–N* curve for human heart compared to metal *N* fatigue cycle life versus *S* stress amplitude 228
Figure C.2 Simplified body repair 230
Figure C.3 Charge and repair RC model for the human body 233

List of Tables

Table 1.1	Generalized conjugate mechanical work variables	8
Table 1.2	Some state variables	8
Table 1.3	Some common intensive and extensive thermodynamic variables [1, 4]	9
Table 1.4	Common thermodynamic processes	9
Table 1.5	Thermodynamic aging states [1, 2, 4]	14
Table 2.1	Common time series transforms	43
Table 2.2	Four common thermodynamic potential and energy states	62
Table 4.1	Typical constants for stress–time creep law	96
Table 4.2	Damping loss factor examples for certain materials	107
Table 4.3	Cumulative stress test goals: CAST equations	108
Table 5.1	Estimated relative corrosion resistance	123
Table 5.2	Predicted corrosion rates and amounts for 1 year at 5 mA of current for anodic different metals [7]	125
Table 6.1	Failure mechanisms and associated thermal activation energies	140
Table A.1	Constant failure rate conversion table	187
Table A.2	Relationship between Cpk index and yield [1]	198
Table A.3	Life test data arranged for plotting	200
Table A.4	Field data and the renormalized groups	201
Table A.5	Multiple stress accelerated test to demonstrate 1 FMH [1]	205
Table B.1	MTTF observed	212
Table B.2	Machine stress MTTF observed	214
Table B.3	Gaussian probabilities (%)	219
Table B.4	Multiple stress accelerated test to demonstrate 0.6 FMH	221
Table B.5	Optimized multiple stress accelerated test for 0.6 FMH	221
Table B.6	Profile of a product's temperature exposure per year	223
Table C.1	Human cyclic engine and possible stresses that shorten cycle life	227

About the Author

Dr Alec Feinberg received his Ph.D. in physics from Northeastern University in 1981. He is the founder of DfRSoft, a software and consulting company where he now works. He is the principal author of the book *Design for Reliability*. Alec has 30 years of experience in the area of reliability physics working in diverse industries including AT&T Bell Labs, TASC, M/A-COM, and Advanced Energy. He has provided reliability engineering services in all areas that include solar, thin film, power electronics, defense, microelectronics, aerospace, wireless electronics, and automotive electrical systems. He has provided training classes in Design for Reliability, Shock and Vibration, HALT, Reliability Growth, Electrostatic Discharge, Dielectric Breakdown, DFMEA, and Thermodynamic Reliability Engineering. Alec has presented numerous technical papers and won the 2003 RAMS Alan O. Plait best tutorial award for the topic *Thermodynamic Reliability Engineering*. Alec is also a contributing author to the book *The Physics of Degradation in Engineered Materials and Devices*. Alec is available to provide consulting and to give training classes in Thermodynamic Degradation Science, Design for Reliability, and Shock and Vibration through the author's website at www.dfrsoft.com.

Preface

Thermodynamic degradation science is a new and exciting discipline. It contributes to both physics of failure and as a new area in thermodynamics. There are many different ways to approach the science of degradation. Since thermodynamics uses an energy perspective, it is a great way to analyze such problems. There is something in this book for everyone who is concerned with degradation issues. Even if you are just interested in reliability or accelerated testing, there is a lot of new and highly informative material. We also go beyond traditional physics of failure methods and develop conjugate work models and methods. It is important to have new tools such as "mesoscopic" noise degradation measurements for complex systems and a conjugate work approach to solving physics of failure problems. We cover a number of original key topics in this book, including:

- thermodynamic principles of degradation;
- conjugate work, entropy damage, and free energy degradation analysis;
- physics of failure using conjugate work approach;
- complex systems degradation analysis using noise analysis;
- *mesoscopic* noise entropy measurement for disorder in operating systems;
- human heart degradation noise measurements;
- cumulative entropy damage, cyclic work, and fatigue analysis;
- Miner's rule derivation for fatigue and Miner's rule for batteries;
- engines and efficiency degradation;
- aging laws, cumulative accelerated stress test (CAST) plans, and acceleration factors for: creep; wear; fatigue; thermal cycle; vibration (sine and random); temperature; humidity and temperature;
- transistor aging laws (bipolar and FET models);
- new accelerated test environmental profiling CAST planning method;
- vibration cumulative damage (sine and random);
- FDS (fatigue damage spectrum) analysis (sine and random);
- chemical corrosion and activations aging laws;
- diffusion aging laws;

- reliability statistics;
- how aging laws affect reliability distributions;
- human engine degradation;
- human heart versus metal cyclic fatigue;
- human growth and repair model;
- negative entropy and spontaneous negative entropy; and
- environmental degradation and pollution.

When we think of thermodynamic degradation, whether it be for complex systems, devices, or even human aging, we begin to realize that it is all about "order" being converted to "disorder" due to the natural spontaneous tendencies described by the second law of thermodynamics to come to equilibrium with the neighboring environment. Although most people who study thermodynamics are familiar with its second law, not many think of it as a good explanation of why a product degrades over time. However, we can manipulate and rephrase it as follows.

> *The second law in terms of system thermodynamic degradation: the spontaneous irreversible degradation processes that take place in a system interacting with its environment will do so in order to go towards thermodynamic equilibrium with its environment.*

We see that the science presents us with a gift, for its second law actually explains the aging processes. When I first realized this, I started to combine the science of degradation with thermodynamics. I presented these concepts in a number of papers and conferences, and in the book called *Design for Reliability* first published in 2000. The initial work was done with Professor Alan Widom at Northeastern University (1995). Recently I was invited to write a chapter in a book edited with Professor Swingler at Heriot-Watt University, Edinburgh, entitled *The Physics of Degradation in Engineered Materials and Device.* That gave me the chance to start to work on applications and new ways of performing degradation analysis. We see that this science is starting to catch on. This book presents the fundamentals and goes beyond including new ways to make measurements, and provides many examples so the reader will learn the value of how this science can be used. I believe this science will significantly expand soon and it is my hope that this book will provide the spark to inspire others. I believe there are a lot of new opportunities to enhance and use thermodynamic degradation methods. We should find that prognostics, using a thermodynamic energy approach, should advance our capabilities immensely. I have included such a measurement system in the book.

The fact is that, in many situations, failure is simply not an option and it can take immense planning to prevent failure. We need all the tools available to assist us. Thermodynamic degradation science offers new tools, new ways to solve physics of failure problems, and new ways to do prognostics and prevent failure.

1

Equilibrium Thermodynamic Degradation Science

1.1 Introduction to a New Science

Thermodynamic degradation science is a new and exciting discipline. Reviewing the literature, one might note that thermodynamics is underutilized for this area. You may wonder why we need another approach. The answer is: in many cases you do not. However, the depth and pace of understanding physics of failure phenomenon, and the simplified methods it offers for such problems, is greatly improved because thermodynamics offers an energy approach. Further, systems are sometimes complex and made up of many components. How do we describe the aging of a complex system? Here is another possibility where thermodynamics, an energy approach, can be invaluable. We will also see that assessing thermodynamic degradation can be very helpful in quantifying the life of different devices, their aging laws, understanding of their failure mechanisms and help in reliability accelerated test planning [1–4]. One can envision that degradation is associated with some sort of device damage that has occurred. In terms of thermodynamics, degradation is about order versus disorder in the system of interest. Therefore, often we will use the term thermodynamic damage which is associated with disorder and degradation. One clear advantage to this method is that:

> *thermodynamics is an energy approach, often making it easier to track damage due to disorder and the physics of failure of aging processes.*

More importantly, thermodynamics is a natural candidate to use for understanding system aging.

> *Here the term "system" can be a device, a complex assembly, a component, or an area of interest set apart for study.*

Thermodynamic Degradation Science: Physics of Failure, Accelerated Testing, Fatigue, and Reliability Applications, First Edition. Alec Feinberg.

Although most people who study thermodynamics are familiar with its second law, not many think of it as a good explanation of why a system degrades over time. We can manipulate a phrasing of the second law of thermodynamics to clarify our point [1, 4].

Second law in terms of system thermodynamic damage: the spontaneous irreversible damage processes that take place in a system interacting with its environment will do so in order to go towards thermodynamic equilibrium with its environment.

There are many phrasings of the second law. This phrasing describes aging, and we use it in this chapter as the second law in terms of thermodynamics damage occurring in systems as they age. We provide some examples in Chapter 2 (see Sections 2.10 and 2.11) of this statement in regards to aging to help clarify this.

When we state that degradation is irreversible, we mean either non-repairable damage or that we cannot reverse the degradation without at the same time employing some new energetic process to do so. We see there is a strong parallel consequence of the second law of thermodynamics associated with spontaneous degradation processes.

The science presents us with a gift, for its second law actually explains the aging processes [1, 4].

We are therefore compelled to look towards this science to help us in our study of system degradation. Currently the field of physics of failure includes a lot of thermodynamic-type explanations. Currently however, the application of thermodynamics to the field of device degradation is not fully mature. Its first and second laws can be difficult to apply to complex aging problems. However, we anticipate that a thermodynamic approach to aging will be invaluable and provide new and useful tools.

1.2 Categorizing Physics of Failure Mechanisms

When we talk about system damage, we should not lose sight of the fact that that we are using it as an applicable science for physics of failure. To this end, we would like to keep our sights on this goal. Thermodynamic reliability is a term that can apply to thermodynamic degradation physics of a device after it is taken out of the box and subjected to its use under stressful environmental conditions. We can categorize degradation into categories [1, 2] as follows.

The irreversible mechanisms of interest that cause aging are categorized into four main categories of:
forced processes;
activation;
diffusion; and
combinations of these processes yielding complex aging.

These are the key aging mechanisms typically of interest and discussed in this book. Aging depends often on the rate-controlling process. Any one of these three processes may dominate depending on the failure mode. Alternately, the aging rate of each process may be on the same time scale, making all such mechanisms equally important. Figure 1.1 is a conceptualized overview of these processes and related physics of failure mechanisms.

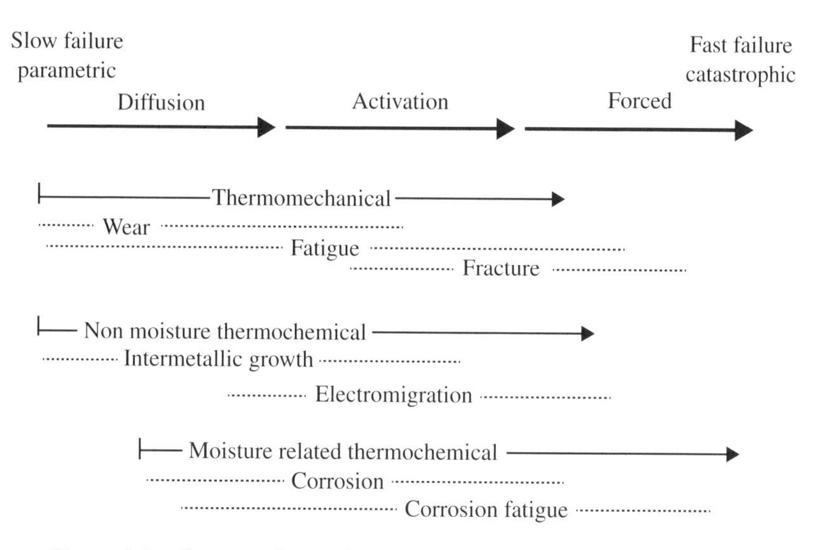

Figure 1.1 Conceptualized aging rates for physics-of-failure mechanisms

In this chapter, we will start by introducing some of the parallels of thermodynamics that can help in our understanding of physics of degradation problems. Here fundamental concepts will be introduced to build a basic framework to aid the reader in understanding the science of thermodynamic damage in physics of failure applications.

1.3 Entropy Damage Concept

When building a semiconductor component, manufacturing a steel beam, or simply inflating a bicycle tire, a system is created which interacts with its environment. Left to itself, the interaction between the system and environment degrade the system of interest in accordance with our second law phrasing of device degradation. Degradation is driven by this tendency of the system/device to come into thermodynamic equilibrium with its environment. The total order of the system plus its environment tends to decrease. Here "order" refers to how matter is organized, for example disorder starts to occur when: the air in the bicycle tire starts to diffuse through the rubber wall; impurities from the environment diffuse into otherwise pure semiconductors; internal manufacturing stresses cause dislocations to move into the semiconductor material; or iron alloy steel beams start to corrode as oxygen atoms from the atmospheric environment diffuse into the steel. In all of these cases, the spontaneous processes creating disorder are irreversible. For example, the air is not expected to go back into the bicycle tire; the semiconductor will not spontaneously purify; and the steel beam will only build up more and more rust. The original order created in a manufactured product diminishes in a random manner, and becomes measurable in our macroscopic world.

Associated with the increase in total disorder or entropy is a loss of ability to perform useful work. The total energy has not been lost but degraded. The total energy of the system plus the environment is conserved during the process when total thermodynamic equilibrium is approached. The entropy of the aging process is associated with that portion of matter that has become disorganized and is affecting the ability of our device to do useful work. For the bicycle tire example, prior to aging the system energy was in a highly organized state. After aging, the energy of the gas molecules (which were inside the bicycle tire) is now randomly distributed in the environment. These molecules cannot easily perform organized work; the steel beam, when corroded into rust, has lost its strength. These typical second-law examples describe the irreversible processes that cause aging.

More precisely:

if entropy damage has not increased, then the system has not aged.

Sometimes it will be helpful to separately talk about entropy in two separate categories.

Entropy damage causes system damage, as compared to an entropy term we refer to as non-damage entropy flow.

For example, the bicycle tire that degraded due to energy loss did not experience damage and can be re-used. Adding heat to a device increased entropy but did not necessarily cause damage. However, the corrosion of the steel beam is permanent damage. In some cases, it will be obvious; in other cases however, we may need to keep tabs on entropy damage. In most cases, we will mainly be looking at entropy change due to device aging as compared to absolute values of entropy since entropy change is easier to measure. Entropy in general is not an easy term to understand. It is like energy: the more we learn how to measure it, the easier it becomes to understand.

1.3.1 The System (Device) and its Environment

In thermodynamics, we see that it is important to define both the device and its neighboring environment. Traditionally, this is done quite a bit in thermodynamics. Note that most books use the term "system" [5, 6]. Here this term applies to some sort of device, complex subsystem, or even a full system comprising many devices. The actual term system or controlled mass is often used in many thermodynamic text books. In terms of the aging framework, we define the following in the most general sense.

- *The system is some sort of volume set apart for study. From an engineering point of view, of concern is the possible aging of the system that can occur.*
- *The environment is the neighboring matter, which interacts with the system in such a way as to drive it towards its thermodynamic equilibrium aging state.*
- *This interaction between a system and its environment drives the system towards a thermodynamic equilibrium lowest-energy aging state.*

It is important to realize that there is no set rule on how the system or the environment is selected. The key is that the final results be consistent.

As a system ages, work is performed by the system on the environment or vice versa. The non-equilibrium process involves an energy exchange between these two.

- *Equilibrium thermodynamics provides methods for describing the initial and final equilibrium system states without describing the details of how the system evolves to a final equilibrium state. Such final states are those of maximum total entropy (for the system plus environment) or minimum free energy (for the system).*
- *Non-equilibrium thermodynamics describes in more detail what happens during the evolution towards the final equilibrium state, for example the precise rate of entropy increase or free energy decrease. Those parts of the energy exchange broken up into heat and work by the first law are also tracked during the evolution to an equilibrium final state. This is a point where the irreversible process virtually slows to a halt.*

1.3.2 Irreversible Thermodynamic Processes Cause Damage

We can elaborate on reversible or irreversible thermodynamic processes. Sanding a piece of wood is an irreversible process that causes damage. We create heat from friction which raises the internal energy of the surface; some of the wood is removed creating highly disordered wood particles so that the entropy has increased. The disordered wood particles can be thought of as entropy damage; the wood block has undergone an increase in its internal energy from heating, which also increases its entropy as well as some of the wood at the surface is loose. Thus, not all the entropy production goes into damage (removal of wood). Since we cannot perform a reversible cycle of sanding that collects the wood particles and puts them back to their original state, the process is irreversible and damage has occurred. Although this is an exaggerated example:

> *in a sense there are no reversible real processes; this is because work is always associated with energy loss.*

The degree of this loss can be minimized in many cases for a quasistatic process (slow varying in time). We are then closer to a reversible process or less irreversible. For example, current flowing through a transistor will cause the component to heat up and emit electromagnetic radiation which cannot be recovered. As well, commonly associated with the energy loss is degradation to the transistor. This is a consequence of the environment performing work on the transistor. In some cases, we could have a device doing work on the environment such as a battery. There are a number of ways to improve the irreversibility of the aging transistor: improve the reliability of the design so that less heat is generated; or lower the environmental stress such as the power applied to the transistor. In the limit of reducing the stress to zero, we approach a reversible process.

> *A reversible process must be quasistatic. However, this does not imply that all quasistatic processes are reversible.*

In addition, the system may be repairable to its original state from a reliability point of view.

> *A quasistatic process ensures that the system will go through a gentle sequence of states such that a number of important thermodynamic parameters are well-defined functions of time; if*

infinitesimally close to equilibrium (so the system remains in quasistatic equilibrium), the process is typically reversible.

A repairable system is in a sense "repairable-reversible" or less irreversible from an aging point of view. However, we cannot change the fact that the entropy of the universe has permanently increased from the original failure and that a new part had to be manufactured for the replaceable part. Such entropy increase has in some sense caused damage to the environment that we live in.

1.4 Thermodynamic Work

As a system ages, work is performed by the system on the environment or vice versa. The non-equilibrium process involves an energy exchange between these two. Measuring the work isothermally (constant temperature) performed by the system on the environment, and if the effect on the system can be quantified, then a measure of the change in the system's free energy can be obtained. If the process is quasistatic, then generally the energy in the system ΔU can be decomposed into the work δW done by the environment on the system and the heat δQ flow.

The bending of a paper clip back and forth illustrates cyclic work done by the environment on the system that often causes dislocations to form in the material. The dislocations cause metal fatigue, and thereby the eventual fracture in the paper clip; the diffusion of contaminants from the environment into the system may represent chemical work done by the environment on the system. We can quantify such changes using the first and second law of thermodynamics. The first law is a statement that energy is *conserved* if one regards heat as a form of energy.

The first law of thermodynamics: the energy change of the system dU is partly due to the heat δQ added to the system which flows from the environment to the system and the work δW performed by the system on the environment (Figure 1.2a):

$$dU = \delta Q - \delta W. \tag{1.1}$$

In the case where heat and work are added to the system, then either one or both can cause damage (Figure 1.2b). If we could track this, we could measure the portion of entropy related to the damage causing the loss in the free energy of the system (which is discussed in Chapter 2).

If heat flows from the system to the environment, then our sign convention is that $\delta Q < 0$. Similarly, if the work is done by the system on the environment then our sign convention is that $\delta W < 0$. That is, adding δQ or δW to the system is positive, increasing the internal energy. In terms of degradation, the first law does not prohibit a degraded system from spontaneous repair which is a consequence of the second law.

Because work and heat are functions of how they are performed (often termed "path dependent" in thermodynamics), we use the notation δW and δQ (for an imperfect differential) instead of dW and dQ, denoting this for an infinitesimal increment of work and heat done along a specific work path or way of adding heat $W_{21} = \int_1^2 \delta W$. We see however that the internal energy is not path dependent, but only depends on the initial and final states of the system so that $\Delta U = U_{\text{final}} - U_{\text{initial}}$. The internal energy is related to all the microscopic forms of energy

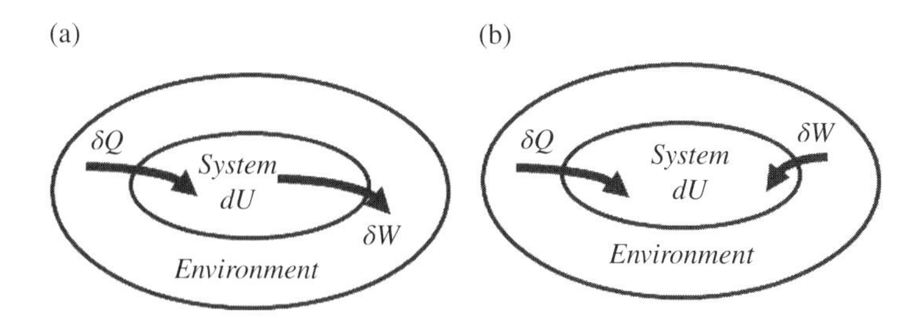

Figure 1.2 First law energy flow to system: (a) heat-in, work-out; and (b) heat-in and work-in

of a system. It is viewed as the sum of the kinetic and potential energies of the molecules of the system if the system is subject to motion, and can be broken down as $\Delta E = \Delta U + \Delta \mathrm{PE} + \Delta \mathrm{KE}$. In most situations, the system is stationary and $\Delta \mathrm{PE} = \Delta \mathrm{KE} = 0$ so that $\Delta U = \Delta E$. Furthermore, in the steady state the internal energy is unchanged, often expressed as $dU_{\text{steady-state}}/dt = 0$. When this is the case we can still have $\delta Q/dt$ and $\delta W/dt$ non-zero. For example, this can occur with a quasistatic heat engine, if heat enters into the system and work is performed by the system, in accordance with the first law, leaving the system unchanged during the process, then the internal energy $dU/dt = 0$. However, if damage occurs to the system during a process, then disorder will occur in the system and the internal energy will of course change.

During the quasistatic process, the work done on the system by the environment has the form

$$\delta W = \sum_a Y_a dX_a. \tag{1.2}$$

Each generalized displacement dX_a is accompanied by a generalized conjugate force Y_a. For a simple system there is but one displacement X accompanied by one conjugate force Y. Key examples of basic conjugate work variables are given in Table 1.1 [1, 4].

1.5 Thermodynamic State Variables and their Characteristics

A system's state is often defined by macroscopic state variables. This can be at a specific time or, more commonly, we use thermodynamic state variables to define the equilibrium state of the system [5, 6]. The equilibrium state means that the system stays in that state usually over a long time period. Common equilibrium states are thermal (when the temperature is the same throughout the system), mechanical (when there is no movement throughout the system), and chemical (the chemical composition is settled and does not change). Common examples of state variables are temperature, volume, pressure, energy, entropy, number of particles, mass, and chemical composition (Table 1.2).

We see that some pairs of state variables are directly related to mechanical work in Table 1.1. These macroscopic parameters depend on the particular system under study and can include voltage, current, electric field, vibration displacements, and so forth.

Thermodynamic parameters can be categorized as intensive or extensive [5, 6]. Intensive variables have uniform values throughout the system such as pressure or temperature. Extensive variables are additive such as volume or mass. For example, if the system is sectioned into

Table 1.1 Generalized conjugate mechanical work variables

Common systems δW	Generalized force (intensive) Y	Generalized displacement (extensive) X	Mechanical work $\delta W = YdX$
Gas	Pressure ($-P$)	Volume (V)	$-P\,dV$
Chemical	Chemical potential (μ)	Molar number of atoms or molecules (N)	$\mu\,dN$
Spring	Force (F)	Distance (x)	$F\,dx$
Mechanical wire/bar	Tension (J)	Length (L)	$J\,dL$
Mechanical strain	Stress (σ)	Strain (e)	$\sigma\,de$
Electric polarization	Polarization (p)	Electric field (E)	$-p\,de$
Capacitance	Voltage (V)	Charge (q)	$V\,dq$
Induction	Current (I)	Magnetic flux (Φ)	$I\,d\Phi$
Magnetic polarizability	Magnetic intensity (H)	Magnetization (M)	$H\,dM$
Linear system	Velocity (v)	Momentum (m)	$v\,dm$
Rotating fluids	Angular velocity (ω)	Angular momentum (L)	$\omega\,dL$
General electrical resistive	Power (voltage $V \times$ current I)	Time (t)	$VI\,dt$

Source: Feinberg and Widom [1], reproduced with permissions of IEEE; Feinberg and Widom [2], reproduced with permission of IEST.

Table 1.2 Some state variables

State variables	Symbol
Pressure	P
Temperature ($dT = 0$ isothermal)	T
Volume	V
Energy (internal, Helmholtz, Gibbs, enthalpy)	U
Entropy	S
Mass, number of particles	m, N
Charge	q
Chemical potential	μ
Chemical composition	X
System noise (Chapter 2)	σ

two subsystems, the total volume V is equal to the sum of the volumes of the two subsystems. The pressure is intensive. The intensive pressures of the subsystems are equal and the same as before the division. Extensive variables become intensive when we characterize them per unit volume or per unit mass. For example, mass is extensive, but density (mass per unit volume) is intensive. Some intensive parameters can be defined for different materials such as density, resistivity, hardness, and specific heat. Conjugate work variables δW are made up of intensive Y and extensive dX pairs. Table 1.3 is a more complete list of common intensive and extensive variables.

Table 1.3 Some common intensive and extensive thermodynamic variables [1, 4]

Intensive (non-additive)	Symbol	Extensive (additive)	Symbol
Pressure	P	Volume	V
Temperature	T	Entropy	S
Chemical potential	μ	Particle number	N
Voltage	V	Electric charge	Q
Density	ρ	Mass	M
Electric field	E	Polarization	P
Specific heat (capacity per unit mass C, at constant volume, pressure, $c_p = c_v$ for solids and liquids), $c = C/m$	c_v, c_p	Heat capacity C_p, C_v of a substance $C = \delta Q/dT$	C_p, C_v
Magnetic field	M	Heat	Q
Compressibility (adiabatic)	β_K	Length, area	L, A
Compressibility (isothermal)	β_T	Internal energy	U
Hardness	h	Gibbs free energy	G
Elasticity (Young's modulus)	Y	Helmholtz free energy	F, A
Boiling or freezing point	T_B, T_F	Enthalpy	H
Electrical resistivity	ζ	Resistance	R

Source: Feinberg and Widom [1], reproduced with permissions of IEEE; adapted from Swingler [4].

Table 1.4 Common thermodynamic processes

Process	Definition
Isentropic	Adiabatic process where $\delta Q = 0$ and the work transferred is frictionless and is therefore reversible
Adiabatic	A process that occurs without transfer of heat $\delta Q = 0$ or mass $dm = 0$, energy is transferred only as work
Isothermal	A change to the system where temperature remains constant, that is, $dT = 0$, typically when the system is in contact with a heat reservoir
Isobaric	A process taking place at constant pressure, $dP = 0$
Isochoric	A process taking place at constant volume, $dV = 0$
Quasistatic	A gentle process occurring very slowly so the system remains in internal equilibrium; any reversible process is quasistatic
Reversible	An idealized process that does not create entropy
Irreversible	A process that is not reversible and therefore creates entropy

Lastly, Table 1.4 provides a table with definitions of common thermodynamic processes that can be helpful as a quick reference.

1.6 Thermodynamic Second Law in Terms of System Entropy Damage

We have stated that as a device ages, measurable disorder (degradation) occurs. We mentioned that the quantity entropy defines the property of matter that measures the degree of microscopic or *mesoscopic* disorder that appears at the macroscopic level. (Microscopic and mesoscopic disorder is better defined in Chapter 2.) We can also re-state our phrasing for system aging of the second law in terms of entropy when the system–environment is isolated.

> *Second law in entropy terms of system degradation: the spontaneous irreversible damage process that takes place in the system–environment interaction when left to itself increases the entropy damage which results from the tendency of the system to go towards thermodynamic equilibrium with its environment in a measurable or predictable way, such that:*

$$\Delta S_{\text{damage}} \geq 0. \tag{1.3}$$

This is the second law of thermodynamics stated in terms of device damage. *We use the term "entropy damage" which is unique to this book; we define it here as we feel it merits special attention* [1, 2]. Some authors refer to it as internal irreversibility [7, 8]. We like to think of it as something that is measurable or predictable, so we try to differentiate it (see the following section). For example, we discuss the fatigue limit in Section 1.6.1 (see Figure 1.3) which necessitates an actual entropy damage threshold. Once entropy damage occurs, there is no way to reverse this process without creating more entropy. This alternate phrasing of the second law is another way of saying that the total order in the system plus the environment changes towards disorder. Degradation is a natural process that starts in one equilibrium state and ends in another; it will go in the direction that causes the entropy of the system plus the environment to increase for an irreversible process, and to remain constant for a reversible process. If we consider the system and its surroundings as an isolated system (where there is no heat transfer in or out), then the entropy of an isolated system during any process always increases (see also Equation (1.7) with $\delta Q = 0$) or, in the limiting case of a reversible process, remains constant, so that

$$\Delta S_{\text{isolated}} \geq 0. \tag{1.4}$$

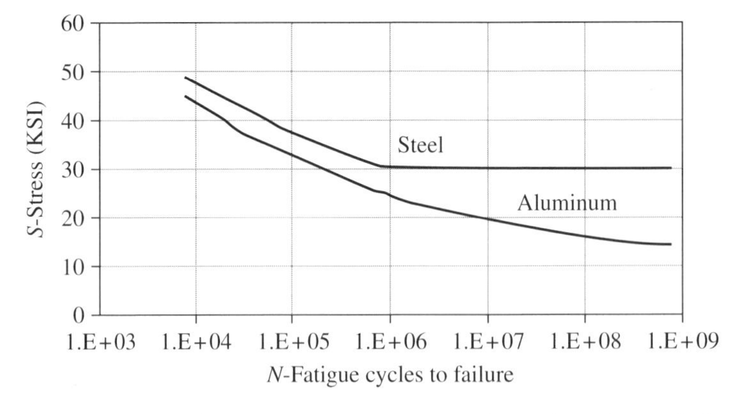

Figure 1.3 Fatigue *S–N* curve of cycles to failure versus stress, illustrating a fatigue limit in steel and no apparent limit in aluminum

1.6.1 Thermodynamic Entropy Damage Axiom

Entropy is an extensive property, so that the total entropy between the environment and the system is the sum of the entropies of each. The system and its local environment can therefore be isolated to help explain the entropy change. We can write that the entropy generated S_{gen} in any isolated process as an extension of Equation (1.4):

$$S_{gen\text{-}any\text{-}process} = \Delta S_{total} = \Delta S_{system} + \Delta S_{env} \geq 0. \tag{1.5}$$

For any process, the overall exchange between an isolated system and its neighboring environment requires that the entropy generated be greater than zero for an irreversible process and equal to zero for a reversible process.

This is also a statement of the second law of thermodynamics. However, since no actual real process is truly reversible, we know that some entropy increase is generated; we therefore have the expression that the entropy of the universe is always increasing (which is a concern because it is adding disorder to our universe). However, if the system is not isolated and heat or mass can flow in or out, then entropy of the system can in fact increase or decrease. This leads us to the requirement of defining entropy damage and non-damage entropy flow related to the system.

In a degradation process, the system and the environment can both have their entropies changed, for example matter that has become disorganized such as a phase change that affects device performance. Not all changes in system entropy cause damage. In theory, "entropy damage" is separable in the aging process related to the device (system) such that we can write the entropy of the system:

$$\Delta S_{system} = \Delta S_{damage} + \Delta S_{non\text{-}damage}; \quad \Delta S_{damage} \geq 0 \tag{1.6}$$

where $\Delta S = S_{in} - S_{out}$ or $\Delta S = S_{initial} - S_{final}$. Combining equations (1.6) and (1.5), we can write

$$S_{gen\text{-}any\text{-}process} = \Delta S_{total} = \Delta S_{damage} + \Delta S_{non\text{-}damage} + \Delta S_{env} \geq 0. \tag{1.7}$$

We require that entropy damage is always equal to or greater than zero, as indicated by Equation (1.3). However, $\Delta S_{non\text{-}damage}$ may or may not be greater than zero. The term $\Delta S_{non\text{-}damage}$ is often referred to by other authors [7, 8] as entropy exchanged or entropy flow with the environment. If the change in the non-damage entropy flow leaves the system and is greater than the entropy damage, then the system can actually become more ordered. However, we still have the entropy damage which can only increase and create failure.

Entropy can be added by adding mass for example, which typically does not cause degradation. Measurements to track entropy damage that generate irreversibilities in a system can be helpful to predict impending failure. The non-damage entropy flow term helps us track the other thermodynamic processes occurring and their direction.

Damage can also have stress thresholds in certain materials. For example, fatigue testing has shown that some materials exhibit a fatigue limit where fatigue ceases to occur below a limit, while other materials have no fatigue limit. If there is damage occurring in materials below their fatigue limit, it is outside our measurement capability; it is therefore not of practical concern and we are not inclined to include it as entropy damage. Figure 1.3 illustrates what is called an

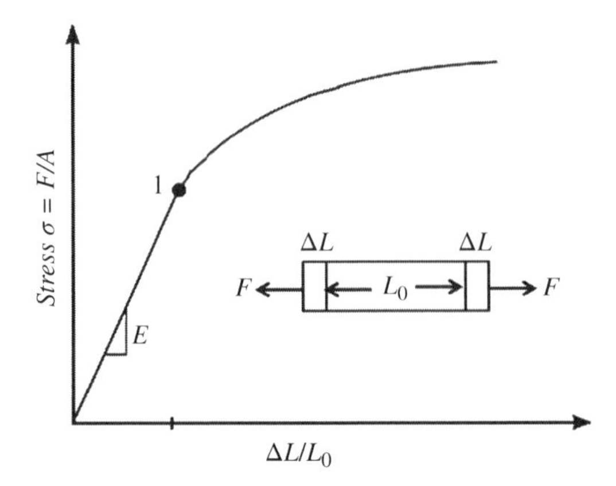

Figure 1.4 Elastic stress limit and yielding point 1

S–N curve for stress *S* versus number of cycles *N* to failure for steel and aluminum. We note that steel has an apparent fatigue limit at about 30 ksi, below which entropy damage is not measureable. Although there is resistance in the restoring force, meaning that heat is likely given off, damage is not measurable.

We can think of the fatigue limit as an entropy damage threshold.

Entropy damage threshold occurs when the entropy change in the system is not significant enough to cause measurable system damage until we go above a certain threshold.

Another example of a subtle limit is prior to yielding, there is an elastic range as shown in Figure 1.4 at point 1. In the elastic range, no measurable damage occurs. For example, prior to creep, stress and strain can be thought of as a quasistatic process that is approximately reversible. Stress σ_e (=force/area) below the elastic limit will cause a strain $\Delta L/L_0$ that is proportional to Young's Modulus E:

$$\frac{\Delta L}{L_0} = \frac{\sigma_e}{E}.$$

Above this elastic limit, yielding occurs. Once the strain is removed in the elastic region, the material returns to its original elongation L_0 so that $\Delta L = 0$ and no measurable entropy damage was observed. In alternate thermodynamic language it might be asserted that, since no process of this sort is in fact reversible, some damage occurred. In this book we would like to depart slightly from that convention and look at practical measurable change rather than be bothered by semantics. If change is not measurable or predictable, then entropy damage can be considered negligible.

Non-damage entropy flow increase $\Delta S_{\text{non - damage}}$ can occur in many system processes, and may be thought of as more disorganization occurring in the system that is not currently affecting the system performance or its energy is eventually exchanged to the local environment.

Entropy Damage Axiom: While non-damage entropy flow can be added or removed from a system without causing measurable and/or easily predictable degradation, such as by adding or removing heat in the system, damage system entropy can only increase in a likely measurable or predictable way to its maximum value where the system approaches failure.

We have now clarified what we mean by entropy damage. We recognize the fact that system entropy increase can occur and we might not have the capability to measure it or at least predict it. However, defining entropy damage in this way presents us with the challenge of actually finding a way to measure or predict what is occurring; after all, that is an important goal.

1.6.2 Entropy and Free Energy

Prior to aging, our system has a certain portion of its energy that is "available" to do useful work. This is called the thermodynamic *free energy*. The thermodynamic free energy is the internal energy input of a system minus the energy that cannot be used to create work. This unusable energy is in our case the entropy damage S multiplied by the temperature T of the system. At this point we may sometimes drop the notation for entropy damage when it is clear that system degradation will refer to an increase in entropy damage and a decrease in the free energy of the system due to entropy damage. It is important to keep this in mind as, in some thermodynamic systems, the free energy decrease is due to a loss of energy such as fuel or work done by heat; the system free energy decrease is therefore not entirely due to system aging. Here we need to rethink our definitions in terms of system degradation. With that in mind, we discuss the common free energies used and later in Chapter 2 (Section 2.11) provide an example relating to system degradation measurements.

There are two thermodynamic free energies widely used: (1) the Helmholtz free energy F, which is the capacity to do mainly mechanical (useful) work (used at constant temperature and volume); and (2) the Gibbs' energy which is primarily for non-mechanical work in chemistry (used at constant temperature and pressure). The system's free (or available) energy is in practice less than the system energy U; that is, if T denotes the temperature of the environment and S denotes the system entropy, then the Helmholtz free energy is $F = U - TS$, which obeys $F < U$ (see Section 2.11.1). The free energy is then the internal energy of a system minus the amount of unusable (or useless) energy that cannot be used to perform work. In terms of system degradation, we can rethink the free energy definition in terms of the entropy damage axiom. That is, we might write the free energy change due to degradation as $\Delta F_{\text{damage}} = \Delta U - T\Delta S_{\text{damage}}$. The unusable energy related to degradation is given by the entropy damage change of a system multiplied by the temperature of the system. If the system's initial free energy is denoted F_{i} (before aging) and the final free energy is denoted F_{f} (after aging), then $F_{\text{f}} < F_{\text{i}}$. The system is in thermal equilibrium with the environment when the free energy is minimized. We can also phrase the second law for aging in terms of the free energy as follows.

Second law in free energy terms of system degradation: the spontaneous irreversible damage process that takes place over time in a system decreases the free energy of the system toward a minimum value. This spontaneous process reduces the ability of the system to perform useful work on the environment, which results in aging of the system.

The work that can be done on or by the system is then bounded by the system's free energy (see Section 2.11.5).

Table 1.5 Thermodynamic aging states [1, 2, 4]

Entropy damage* definitions	Measurement damage definition	Free energy definitions
Non-equilibrium aging		
$\frac{dS_{\text{damage}}}{dt} > 0$	$\frac{d\Delta S_{\text{damage}}}{dt} > 0$	$\frac{d\phi}{dt} < 0$
Equilibrium non-aging state (associated with catastrophic failure)		
$\frac{dS_{\text{total}}}{dt} = 0$	$\frac{d\Delta S_{\text{total}}}{dt} = 0$	$\frac{d\phi}{dt} = 0$
Parametric equilibrium aging state (associated with parametric failure)		
$S_{\text{damage}} \geq S_{\text{para-threshold}}$	$\Delta S_{\text{total}} \geq \Delta S_{\text{threshold}}$	$\phi \leq \phi_{\text{threshold}}$
Relative equilibrium aging state (associated with activated failure)		
$S_{\text{B total}} \leq S_{\text{total}} \leq S_{\text{B total}} + \Delta s$	$\Delta S_{\text{B total}} \leq \Delta S_{\text{total}} \leq \Delta S_{\text{B total}} + \Delta s$	$\phi_{\text{B}} \leq \phi \leq \phi_{\text{B}} + \Delta\phi$

Source: Feinberg and Widom [2], reproduced with permission of IEST; Feinberg and Widom [1], reproduced with permissions of IEEE; adapted from Swingler [4].
* The subscript for entropy damage has been dropped.

$$\text{Work} \leq \Delta\text{Free energy of the system.} \tag{1.8}$$

If the system's free energy is at its lowest state, then (1) the system is in equilibrium with the environment, so ΔFree energy $= 0$ *and (2) the entropy is at a maximum value with a maximum amount of disorder or damage created in the system, so* $\Delta S_{\text{damage}} = 0$ *(see Table 1.5).*

Note also that when the work $W = -\Delta$ Free energy of the system, it occurs in a quasistatic reversible process. As the entropy damage increases, the free energy decreases (see Figure 2.20) as does the available work. The relation between the free energy, entropy, and work is described in detail in the next chapter (see Sections 2.10 and 2.11 and examples in Chapter 4).

1.7 Work, Resistance, Generated Entropy, and the Second Law

Work is done against some sort of resistance. Resistance is a type of friction which increases temperature and creates heat and entropy. In the absence of friction, the change in entropy is zero. We think of mechanical friction as the most common example but we can include electrical resistance, air resistance, resistance to a chemical reaction, resistance of heat transfer, etc. Real work processes always have friction generating heat. If heat flows into a system, then the entropy of that system increases. Heat flowing out decreases the entropy of the system. Some of the heat that flows into a system can cause entropy damage, in other words permanent degradation to the system.

Instead of talking about some form of "absolute entropy," measurements are generally made of the change in entropy that takes place in a specific thermodynamic process. For example:

in an isothermal process, the change in system entropy dS is the change in heat δQ divided by the absolute temperature T:

$$dS_{\text{system}} \geq \frac{\delta Q}{T} \tag{1.9}$$

where T is the temperature at the boundary and δQ is the heat exchanged between the system and the environment.

Here the equality holds for reversible processes or equilibrium condition, and the inequality for a spontaneous irreversible process in non-equilibrium (as defined in the next section). Equation (1.9) is a result of what is known as the Clausius inequality:

$$\oint \frac{\delta Q}{T} \leq 0,$$

where

$$\oint \frac{\delta Q}{T} = \int_1^2 \frac{\delta Q}{T} + \int_2^1 \frac{\delta Q}{T} = \int_1^2 \frac{\delta Q}{T} + S_1 - S_2$$

then

$$\int_1^2 \frac{\delta Q}{T} \leq S_1 - S_2,$$

which gives Equation (1.9). Note that because the entropy measurement depends on our observation of heat flow, which is a function of the heat dissipation process (often termed path dependent in thermodynamics), we again use the notation δQ instead of dQ.

The equality in Equation (1.9) for a reversible process is associated with entropy flow, and it is non-damage entropy flow $dS_{\text{non-damage}} = \delta Q / T$ that is the reversible entropy. When the heat flows from the environment to the system it is positive entropy $(\delta Q > 0)$ and when heat flows from the system out to the environment it is negative $(\delta Q < 0)$. We can then replace the inequality sign by including the irreversibility term of what we call here the entropy damage. Instead of an inequality, we can write this as

$$\begin{aligned} dS_{\text{system}} &= dS_{\text{non-damage}} + dS_{\text{damage}} \\ &= \frac{\delta Q}{T} + dS_{\text{damage}} \quad \text{and} \quad dS_{\text{damage}} \geq 0 \text{ is the } \textbf{second law requirement}. \end{aligned} \tag{1.10}$$

This now verifies Equation (1.6), obtained by removing the inequality in Equation (1.9) and adding the extra term of entropy damage. By the way, entropy damage, like work, is path-dependent on how work was performed. This is also true of entropy non-damage. Although we refrain from using too many confusing symbols here, we should in reality be mindful of the path dependence and could be writing dS_{damage} and $dS_{\text{non-damage}}$ as δS_{damage} and $\delta S_{\text{non-damage}}$. Note that in most thermodynamic books the state variable dS is used; dS is not path dependent, that is, it only depends on the initial and final states and not on the path.

1.8 Thermodynamic Catastrophic and Parametric Failure

We have defined the maximum entropy and minimum free energy as equilibrium states. We now use these definitions to formally define failure.

Catastrophic failure occurs when the system's free energy is as small as possible and its entropy is as large as possible, and the system is in its true final thermodynamic equilibrium state with the neighboring environment such that:

$$\frac{d\phi}{dt}=0 \text{ and } \frac{dS_{\text{total}}}{dt}=0. \tag{1.11}$$

Typically, catastrophic failure occurs due to permanent system degradation caused by maximum entropy damage. We note that there are non-catastrophic intermediate equilibrium states that a system can be in with its environment [9]. Some examples might include when an electrical system is turned off or in a standby state, or if we have a secondary battery that has cycled in numerous charge–discharge cycles where each fatigue cycle represents an intermediate degraded state where the battery has not fully degraded to the point where it cannot perform useful work (see Sections 2.10.2 and 5.1.1). Quasistatic measurements and quasi-intermediate states are discussed more in Section 1.3.2. In such cases, the degraded system has not failed as it is not in its "true final equilibrium state" relative to its stress environment. We therefore have to define what we mean by the equilibrium state relative to looking for a maximum in entropy.

We can also envision a situation in which a device such as a transistor or an engine degrades to a point where it can no longer perform at the intended design level. The transistor's power output may have degraded 20%, the engine's efficiency may have degraded 70%. When a parametric threshold is involved, we are likely not at a true final equilibrium state. Therefore, we note that:

parametric failure occurs when the system's free energy and entropy have reached a critical parametric failure threshold value, such that:

$$\phi \le \phi_{\text{threshold}} \text{ and } S_{\text{damage}} \ge S_{\text{para-threshold}}. \tag{1.12}$$

In such cases, parametric failure is also the result of entropy damage.

1.8.1 Equilibrium and Non-Equilibrium Aging States in Terms of the Free Energy or Entropy Change

At this point we would like to define entropy and free energy in equilibrium and non-equilibrium thermodynamics. We have stated that equilibrium thermodynamics provides methods for describing the initial and final equilibrium system states, without describing the details of how the system evolves to final state. On the other hand, non-equilibrium thermodynamics describes in more detail what happens during evolution to the final equilibrium state, for example the precise rate of entropy increase or free energy decrease. Those parts of the energy exchange broken up into heat and work by the first law are also tracked during the evolution to an equilibrium final state. This is a point where the irreversible process virtually slows to a halt.

For example, as work is performed by a chemical cell (a battery with an electromotive force), the cell ages and the free energy decreases. Non-equilibrium thermodynamics describes the evolution which takes place as current passes through the battery, and the final equilibrium state is achieved when the current stops and the battery is dead. "Recharging" can revive a secondary battery. However, this is cyclic work that also degrades the battery's capacity after each cycle.

Table 1.5 summarizes the key aging states [1, 2, 4].

1.9 Repair Entropy

We now have defined system aging in terms of entropy damage increase and/or free energy decrease (loss of ability to do useful work). We might ask: does it make sense to define a negative entropy term called "repair entropy"? After all, there are many repairable systems. In reliability science, both non-repairable and repairable systems are often treated separately. For example, the term "mean time between failure" (MTBF) is used for repairable systems, while "mean time to failure" (MTTF) is used for non-repairable systems. Another reason for defining "repair entropy" is that it will provide insight into the entropy damage process.

When we repair a system, we must perform work and this work is associated with reorganizing entropy for the system that has been damaged. The quantity of repair entropy or reorganization needed (in most cases) is approximately equal to (or greater than) the entropy damage quantity that occurred to the degraded system, that is

$$\left|\Delta S_{\text{repair}}\right| \geq \left|\Delta S_{\text{damage}}\right|. \tag{1.13}$$

In the reference frame of the system we must have that:

$$\Delta S_{\text{damage}} \geq 0,\ \Delta S_{\text{system-repair}} \leq 0, \quad \text{so} \quad \Delta S_{\text{damage}} + \Delta S_{\text{system-repair}} \approx 0. \tag{1.14}$$

That is our understanding of entropy; the damage system under repair will become reorganized and can again do useful work. Therefore, the systems change in entropy has essentially decreased and its free energy has increased again. There is no free ride however; by the second law (Equation (1.5)) the repair process generated at least this same amount of entropy damage or greater to the environment $|\Delta S_{\text{repair}}|$, often in the form of pollution.

Unlike system entropy damage, which we typically think of as a spontaneous aging process or a tendency for the system under stress to come to equilibrium with its neighboring environment, repair entropy is a non-spontaneous process. It is therefore not a typical thermodynamic term; however, it can be. For example, it is a measure of the quantity of heat needed or its equivalent to perform the repair process (see Section 1.8).

1.9.1 Example 1.1: Repair Entropy: Relating Non-Damage Entropy Flow to Entropy Damage

Consider a repair to a system that requires a certain amount of heat $-\delta Q_{\text{repair}}/T$. This is the repair entropy equivalent. A simple example is a failed solder joint. The repair amount of heat to reflow the solder joint is greater than the equivalent entropy damage created. Essentially, we are equating the entropy flow (or non-damage entropy flow) in this particular case to entropy damage, which is another way of looking at Equation (1.13). That is, when entropy damage occurs, it also creates an amount of entropy heat flow. In this exercise, we note that to carry out a repair we need to reverse the process. This helps to illustrate how, in many situations (of course not all situations), heat loss due to entropy damage creates entropy flow to the environment. This portion of non-damage entropy flow can then be related to entropy damage.

The entropy generated in repair is the sum of the environmental entropy and heat needed for repair, given by:

$$dS_{\text{generated}} = dS_{\text{env}} - \left(\frac{\delta Q}{T}\right)_{\text{system-repair}}. \tag{1.15}$$

We note this heat can be related to energy cost to manufacture the repaired part. We see that by the second law (Equation (1.4)) $dS_{\text{generated}} \geq 0$. Therefore, this means that

$$TdS_{\text{env}} \geq (\delta Q)_{\text{system-repair}}. \tag{1.16}$$

This is as suspected that the repair process generates equal or more disorganized energy to the environment than the amount of organized energy needed for the repair process. This is evident in the solder joint example regarding the heat energy repair needed to reflow the solder joint. Sometimes the non-damage entropy flow to the environment is harmless; at other times, it may be very harmful and cause serious pollution issues.

Lastly, we would like to note that Mother Nature is a special case in which spontaneous entropy repair does exist. We have included this in the Special Topic Chapter C called "Negative Entropy and the Perfect Human Engine."

We have now defined many useful terms, concepts, and key equations for system degradation analysis. The following chapter will provide many examples in equilibrium thermodynamics.

Summary

1.1 Introduction to a New Science

- Thermodynamics is an energy approach, often making it easier to track damage due to disorder and the physics of failure of aging processes.
- Second law in terms of system (device) thermodynamic damage: the spontaneous irreversible damage processes that takes place in a system interacting with its environment will do so in order to go towards thermodynamic equilibrium with its environment.

1.2 Categorizing Physics of Failure Mechanisms

The irreversible mechanisms of interest that cause aging are categorized into four main categories of: (1) forced processes; (2) activation; (3) diffusion; (4) combinations of these processes yielding complex aging.

1.3 Entropy Damage Concept

If entropy damage has not increased, then the device/system has not aged.

In terms of degradation processes, entropy can be described in two categories.

Entropy damage is a term used to describe entropy which causes system damage, as compared with non-damage entropy flow.

1.3.1 The System (Device) and its Environment

- The system is some sort of volume set apart for study. From an engineering point of view, the possible aging that can occur to the system is of concern.
- The environment is the neighboring matter which interacts with the system in such a way as to drive it towards its thermodynamic equilibrium aging state.
- Equilibrium thermodynamics provides methods for describing the initial and final equilibrium system states without describing the details of how the system evolves to a final equilibrium state. Such final states are those of maximum total entropy (for the system plus environment) or minimum free energy (for the system).
- Non-equilibrium thermodynamics describes in more detail what happens during the evolution to the final equilibrium state, for example, the precise rate of entropy increase or free energy decrease. Those parts of the energy exchange broken up into heat and work by the first law are also tracked during the evolution to an equilibrium final state. This is a point where the irreversible process virtually slows to a halt.

1.3.2 Irreversible Thermodynamic Processes Cause Damage

In a sense, there are no reversible real processes. This is because work is always associated with energy loss.

A reversible process must be quasistatic. However, this does not imply that all quasistatic processes are reversible.

In addition, the system may be repairable to its original state from a reliability point of view.

A quasistatic process ensures that the system will go through a gentle sequence of states such that a number of important thermodynamic parameters are well-defined functions of time. If infinitesimally close to equilibrium (so the system remains in quasistatic equilibrium), the process is typically reversible.

1.4 Thermodynamic Work

The first law of thermodynamics: the energy change of the system dU is partly due to the heat δQ added to the system which flows from the environment to the system and the work δW performed by the system on the environment (Figure 1.2a):

$$dU = \delta Q - \delta W. \tag{1.1}$$

In the case where heat and work are added to the system, then either one or both can cause damage (Figure 1.2b). If we could track this, we would find that the portion of entropy is related to the damage which causes a loss in the free energy of the system (discussed in Chapter 2).

During the quasistatic process, the work done on the system by the environment is of the form:

$$\delta W = \sum_a Y_a dX_a. \tag{1.2}$$

Each generalized displacement dX_a is accompanied by a generalized conjugate force Y_a. Table 1.1 summarizes the generalized conjugate mechanical work variables.

1.5 Thermodynamic State Variables and their Characteristics

A system's state is often defined by macroscopic state variables. We use thermodynamic state variables to define the equilibrium state of the system. Common examples of state variables are temperature, volume, pressure, energy, entropy, number of particles, mass, and chemical composition.

Thermodynamic parameters can be categorized as intensive or extensive. Intensive variables have uniform values throughout the system, such as pressure or temperature. Extensive variables are additive such as volume or mass.

1.6 Thermodynamic Second Law in Terms of System Entropy Damage

Second Law in entropy terms of system degradation: the spontaneous irreversible damage process that takes place in the system–environment interaction when left to itself increases the entropy damage which results from a tendency of the system to go towards thermodynamic equilibrium with its environment in a measurable or predictable way.

$$\Delta S_{\text{damage}} \geq 0. \tag{1.3}$$

1.6.1 Thermodynamic Entropy Damage Axiom

Entropy damage axiom: the entropy generated associated with system (device) degradation is referred to as "entropy damage."

We can write that the entropy generated S_{gen} in any isolated process as an extension of Equation (1.4):

$$S_{\text{gen-any-process}} = \Delta S_{\text{total}} = \Delta S_{\text{system}} + \Delta S_{\text{env}} \geq 0. \tag{1.5}$$

Here the equal sign indicates a reversible process. However, since no actual process is truly reversible, we know that some entropy is generated during any process, hence the expression that the entropy of the universe is always increasing.

In theory, "entropy damage" is separable in the aging process related to the system such that

$$S_{\text{gen-any-process}} = \Delta S_{\text{total}} = \Delta S_{\text{damage}} + \Delta S_{\text{non-damage}} + \Delta S_{\text{env}} \geq 0. \tag{1.7}$$

Entropy damage can have stress thresholds in certain materials. For example, fatigue testing has shown that some materials exhibit a fatigue limit below which fatigue does not occur.

Entropy damage threshold occurs when the entropy change in the system is not significant enough to cause measurable system damage until it reaches a certain threshold.

Entropy damage in a system can be stated in more detail when involving measurements or predictions as follows.

Entropy damage axiom: while non-damage entropy flow can be added or removed from a system without causing measurable and or easily predictable degradation, such as by adding or removing heat in the system, damage system entropy can only increase in a likely measurable or predictable way to its maximum value where the system approaches failure.

1.6.2 Entropy and Free Energy

Second law in free energy terms of system degradation: The spontaneous irreversible damage process that takes place over time in a system decreases the free energy of the system toward a minimum value. This spontaneous process reduces the ability of the system to perform useful work on the environment, which results in aging of the system.

The work that can be done on or by the system is then bounded by the system's free energy (see Section 2.11.5):

$$\text{Work} \leq \Delta\text{Free energy of the system.} \tag{1.8}$$

If the system's free energy is at its lowest state, then (1) the system is in equilibrium with the environment, so ΔFree energy = 0; and (2) entropy is at a maximum value with a maximum amount of disorder or damage created in the system, so $\Delta S_{\text{damage}} = 0$ (see Table 1.4).

1.7 Work, Resistance, Generated Entropy, and the Second Law

In an isothermal process, the change in system entropy ΔS is the change in heat δQ divided by the absolute temperature T:

$$dS_{\text{system}} \geq \frac{\delta Q}{T}. \tag{1.9}$$

Instead of the inequality, we can write this as

$$dS_{\text{system}} = dS_{\text{non-damage}} + dS_{\text{damage}} = \frac{\delta Q}{T} + dS_{\text{damage}} \quad \text{where} \quad dS_{\text{damage}} \geq 0. \tag{1.10}$$

1.8 Thermodynamic Catastrophic and Parametric Failure

Catastrophic failure occurs when the system's free energy is as small as possible and its entropy is as large as possible, and the system is in its true final thermodynamic equilibrium state with the neighboring environment such that:

$$\frac{d\phi}{dt}=0 \text{ and } \frac{dS_{\text{total}}}{dt}=0. \tag{1.11}$$

Parametric failure occurs when the system's free energy and entropy have reached a critical parametric failure threshold value such that

$$\phi \le \phi_{\text{threshold}} \text{ and } S_{\text{damage}} \ge S_{\text{para-threshold}}. \tag{1.12}$$

1.8.1 Equilibrium and Non-Equilibrium Aging States in Terms of the Free Energy or Entropy Change

The thermodynamic aging states are summarized in Table 1.5.

1.9 Repair Entropy

When we repair a system, we must do work and this work is associated with reorganizing entropy for the system that has been damaged. The quantity of entropy repair is approximately equal to the entropy damage quantity that occurred to the degraded system, that is

$$\left|\Delta S_{\text{repair}}\right| \ge \left|\Delta S_{\text{damage}}\right|. \tag{1.13}$$

For a system requiring a certain amount of equivalent heat energy for repair we find that

$$TdS_{\text{env}} \ge (\delta Q)_{\text{system-repair}}. \tag{1.16}$$

That is, the repair process generates equal or more disorganized energy to the environment than the amount of organized energy needed for the repair process.

References

[1] Feinberg, A. and Widom, A. (2000) On thermodynamic reliability engineering. *IEEE Transaction on Reliability*, **49** (2), 136.

[2] Feinberg, A. and Widom, A. (1995) Aspects of the thermodynamic aging process in reliability physics. *Institute of Environmental Sciences Proceedings*, **41**, 49.

[3] Feinberg, A., Crow, D. (eds) *Design for Reliability*, M/A-COM 2000, CRC Press, Boca Raton, 2001.

[4] Feinberg, A. (2015) Thermodynamic damage within physics of degradation, in *The Physics of Degradation in Engineered Materials and Devices* (ed J. Swingler), Momentum Press, New York.

[5] Reynolds, W.C. and Perkins, H.C. (1977) *Engineering Thermodynamics*, McGraw-Hill, New York.

[6] Cengel, Y.A. and Boles, M.A. (2008) *Thermodynamics: An Engineering Approach*, McGraw-Hill, Boston.

[7] Khonsari, M.M. and Amiri, M. (2013) *Introduction to Thermodynamics of Mechanical Fatigue*, CRC Press, Boca Raton.

[8] Bryant, M. (2015) Thermodynamics of ageing and degradation in engineering devices and machines, in *The Physics of Degradation in Engineered Materials and Devices* (ed J. Swingler), Momentum Press, New York.

[9] Feinberg, A. and Widom, A. (1996) Connecting parametric aging to catastrophic failure through thermodynamics. *IEEE Transactions on Reliability*, **45** (1), 28.

2

Applications of Equilibrium Thermodynamic Degradation to Complex and Simple Systems: Entropy Damage, Vibration, Temperature, Noise Analysis, and Thermodynamic Potentials

2.1 Cumulative Entropy Damage Approach in Physics of Failure

Entropy is an extensive property. If we isolate an area enclosing the system and its environment such that no heat, mass flows, or work flows in or out, then we can keep tabs on the total entropy (Figure 2.1).

In this case the entropy generated from the isolated area has

> *the total entropy of a system is equal to the sum of the entropies of the subsystems (or parts), so the entropy change is*

$$\Delta S_{\text{system}} = \sum_{i}^{N} \left(\Delta S_{\text{subsystem}}\right)_i. \tag{2.1}$$

We now can define here the Cumulative Entropy Damage Equation:

$$\Delta S_{\text{system-cum-damage}} = \sum_{i}^{N} \left(\Delta S_{\text{subsystem-damage}}\right)_i \tag{2.2}$$

Thermodynamic Degradation Science: Physics of Failure, Accelerated Testing, Fatigue, and Reliability Applications, First Edition. Alec Feinberg.

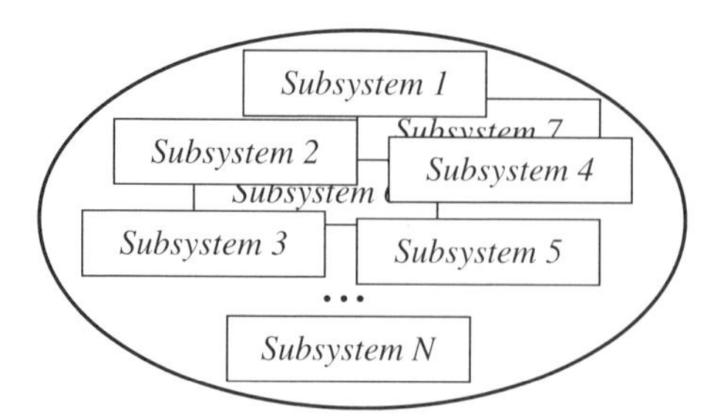

Figure 2.1 The entropy change of an isolated system is the sum of the entropy changes of its components, and is never less than zero

where, for the *i*th subsystem (see Equation (1.9)),

$$\Delta S_{\text{subsystem},i} = \Delta S_{\text{damage},i} + \Delta S_{\text{non-damage},i} \quad \text{and} \quad \Delta \text{S}_{\text{damage},i} \geq 0.$$

This seemingly obvious outcome (Equation (2.2)*) is an important result not only for complex system reliability, but also for device reliability. This will be detailed throughout this chapter.*

Note the non-damage entropy flow can go in or out depending on the process so it is possible for a subsystem's entropy to decrease (become more ordered).

Often in reliability (when redundancy is not used), similar to Equation (2.2), the failure rate λ of the system is equal to the sum of the failure rate of its parts: $\lambda_{\text{system}} = \sum_i^N (\lambda_{\text{subsystem}})_i$.

In this case when any part fails, the system fails. Given that an *i*th subsystem's entropy damage is increasing at a higher rate compared to the other subsystems, then it is dominating system-level damage. Therefore, the impeding failure of the *i*th subsystem is a rate-controlling failure of the system, and its rate of damage affects the system total entropy such that

$$\frac{dS_{\text{system}}}{dt} \approx \frac{dS_i}{dt}. \tag{2.3}$$

If the resolution is reasonable, we can keep tabs on ΔS_{system} damage over time by monitoring the rate controlling subsystem's entropy damage. As long as we can somehow resolve the damage subsystem problem, that is,

$$i\text{th entropy damage resolution} = \frac{\Delta S_{i,\text{damage}}}{\Delta S_{\text{system}}}. \tag{2.4}$$

Sensitivity of our measurement system for determining impending failure may likely depend on a part's entropy damage resolution. This will be more obvious when we are describing noise analysis in Sections 2.5–2.9 and we are concerned about signal to noise issues. In this case, it is the noise rather than the signal that we will be looking at.

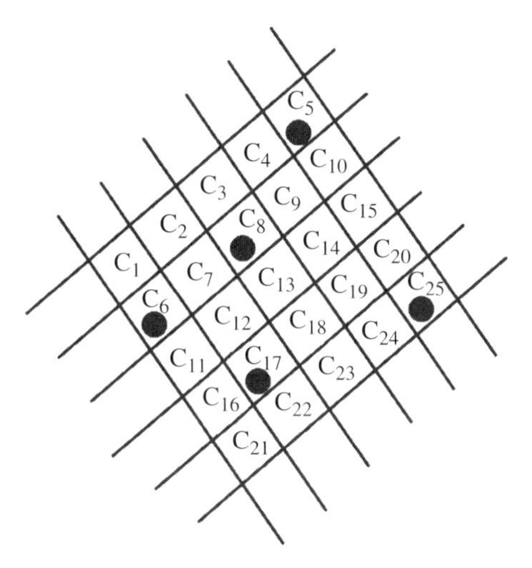

Figure 2.2 Cell fatigue dislocations and cumulating entropy

2.1.1 *Example 2.1: Miner's Rule Derivation*

One basic example of how the important result of Equation (2.2) can be used is to help explain what is termed Miner's fatigue rule, which is a key area in physics of failure.

Consider a subsystem that is made up of, say, metal with defect sites such as a cell lattice shown in Figure 2.2. Just as the entropy of each subsystem is the sum of the entropies, so too is the entropy damage for each cell C which adds up for a subsystem that is under some sort of stress such as cyclic fatigue, as shown in Figure 2.2. Suppose we have n damaged cells so the total entropy damage is

$$\Delta S_{\text{cum-damage-subsystem}} = \sum_{i=1}^{n} \Delta S_{\text{damage-cells},i} \geq 0, \tag{2.5}$$

where n is the total number of cells with entropy damage $S_{\text{damage-cell},i.}$ We assume equal entropy damage per cell. Consider that we have M total cells of potential damage to failure that would lead to a maximum cumulative damage $S_{\max}$. Then the cumulative entropy damage can be normalized:

$$\text{Cum Damage} = \frac{\sum_{n} s_{n,\text{damaged}}}{S_{\max}} = \frac{\sum_{m} m_{\text{damaged-cells}}}{M} \leq 1 \tag{2.6}$$

where s_n is the entropy damage of the nth cell, and m represents a damaged cell such that Σm is the number of damaged cells observed.

If all the cells were damaged the entropy would be at a maximum, then the damage is equal to 1 and this should be equal to counting the number of damaged cells and dividing by the total number of cells that could cause failure. This is one way to measure damage. Here damage is assessed as a value between 0 and 1. Let's assume that there is a linear relationship between the number of cells damaged and the number of fatigue cycles. Fatigue is also dependent on the stress of the cycle, for example, if we are bending metal back and forth, the stress level would be equal to the size of the bend. So for any ith stress level the number of cells damaged for n cycles is some stress fraction f_i of the number of cycles:

$$m_i = f_i n_i \text{ and } M_i = f_i N_i. \tag{2.7}$$

Note that N is the number of cycles to failure and n is the number of cycles performed ($N > n$). Then we can write for the ith stress level:

$$\text{Damage} = \frac{f_1 n_1}{f_1 N_1} + \frac{f_2 n_2}{f_2 N_2} + \ldots + \frac{f_k n_k}{f_k N_k} = \frac{n_1}{N_1} + \frac{n_2}{N_2} + \ldots + \frac{n_k}{N_k} = \sum_{i=1}^{K} \frac{n_i}{N_i}. \tag{2.8}$$

When Damage = 1, failure occurs. This equation is called Miner's rule [1]. Its accuracy then depends on the assumption of Equation (2.7). Furthermore, it follows that we can also invent here a related term called

$$\text{Cumulative Strength} = 1 - \text{Damage}. \tag{2.9}$$

When Damage = 1, the cumulative strength vanishes. Equation (2.7) turns out to be a good approximation, but it is of course not perfectly accurate. We do know that Miner's rule is widely used and has useful validity. Miner's rule is not just applicable to metal fatigue; it can be used for most systems where cumulative damage occurs. Later we have an example in Chapter 5 for secondary batteries. We have now derived it here using the cumulative entropy principle (Equation (2.2)). In Chapter 4 we provide a more accurate way to assess fatigue damage through the thermodynamic work principle.

2.1.2 *Example 2.2: Miner's Rule Example*

Aluminum alloy has the following fatigue characteristics:

Stress 1, $N = 45$ cycles,
Stress 2, $N = 310$ cycles, and
Stress 3, $N = 12\,400$ cycles.

How many times can the following sequence be repeated?

$n_1 = 5$ cycles at stress 1,
$n_2 = 60$ cycles at stress 2, and
$n_3 = 495$ cycles at stress 3

Solution

The fractions of life exhausted in each cycle type are:

$$\frac{n_1}{N_1} = \frac{5}{45} = 0.111, \ \frac{n_2}{N_2} = 0.194\ , \frac{n_2}{N_2} = 0.04.$$

The fraction of life exhausted in a complete sequence yields an approximate cumulative damage of 0.345 (with the cumulative strength decreasing down to 0.655). The life is entirely exhausted when damage $D = 1$ and has been produced 3 times when $D = 1.035$. The sequence can then be repeated about 2.9 times.

One interesting thing to note in fatigue damage is that even though damage accumulates with stress cycles, the degradation may not be easily detectable as the metal system is capable of performing its intended function as long as the system's cumulative strength (Equation (2.9)) is greater than the stress that it is under. We typically find catastrophic failure occurs in a relatively short cycle time compared with the system's cycle life. How then can we detect impending failure at such a *mesoscopic* level (*mesoscopic* is defined in Section 2.5)? One possible solution may be in what we term noise detection. This is discussed later in this chapter (see Sections 2.5–2.9).

2.1.3 Non-Cyclic Applications of Cumulative Damage

The thermodynamic second law requirement that

$$\Delta S_{\text{damage}} \geq 0$$

leads one to realize that even non-cyclic processes must cumulate damage. That is, we can write similar to Miner's Rule, cumulative damage for non-cyclic time-dependent processes

$$\text{Cum Damage} = \sum_{i=1}^{K} \left(\frac{t_i}{\tau_i} \right)^P \tag{2.10}$$

where, by analogy with Miner's cyclic rule, t_i is the time of exposure of the ith stress level and τ_i is the total time to cause failure at the ith stress level. Not all processes age linearly with time as is found in Chapter 4, so the exponent P has been added for the generalized case where P is different from unity (see Chapter 4 for examples and Table 4.3). Equation (2.10) could be used for processes such as creep $(0 < P < 1)$ and wear $(P = 1)$ or even semiconductor degradation, any process where it might make sense to estimate the cumulative damage and/or the cumulative strength left (i.e., 1–Cum Damage). Note the time to failure does not have to be defined as a catastrophic failure. It may be prudent to apply this to parametric failure where τ_i then represents a parametric threshold time for failure. For example, τ_i could be the time when a certain strength, such as voltage output of a battery, degrades by 15%.

The main difference between a cyclic and non-cyclic process is that for a closed system undergoing a quasistatic cyclic process, the internal energy ΔU is often treated theoretically as unchanged in a thermodynamic cycle (see Section 3.3). This is a reversible assumption.

2.2 Measuring Entropy Damage Processes

To make degradation measurements, we need a repeatable method or process to make aging measurements at different times [2]. If we find that the entropy has changed over time from a repeatable quasistatic measurement process, then we are able to measure and track the aging that occurs between the system's initial, intermediate, and final states. We can call this the entropy of an aging process. (Note that during system aging, we do not have to isolate the system. We only need to do this during our measurement process.)

We can theorize that any irreversible process that creates a change in entropy in a system under investigation can cause some degradation to the system. However, if we cannot measure this degradation, then in our macroscopic world the system has not actually aged. In terms of entropy generated from an initial and final state we have:

$$S_{gen} = S_{initial} - S_{final} \geq 0, \tag{2.11}$$

where the equal sign is for a reversible process and the inequality is for an irreversible process. However, what portion of the entropy generated causes degradation to the system and what portion does not? To clarify, similar to Equation (1.4) we have:

$$S_{gen} = S_{damage} + S_{non\text{-}damage}.$$

There is really no easy way to tell unless we can associate the degradation through a measurable quantity. Therefore, in thermodynamic damage, we are forced to define S_{damage} in some measurable way.

As well in thermodynamics, we typically do not measure absolute values of entropy, only entropy change. Let us devise a nearly reversible quasistatic measurement process *f*, and take an entropy measurement of interest at time t_1. Then:

$$\Delta S_f(t_1) = S(t_1 + \Delta t) - S(t_1). \tag{2.12}$$

The measurement process *f* must be consistent to a point that it is repeatable at a much later aging time t_2. If some measurable degradation has occurred to our device, we can observe and record the entropy change:

$$\Delta S_f(t_2) = S(t_2 + \Delta t) - S(t_2), \quad \text{where } t_2 >> t. \tag{2.13}$$

Then we can determine if damage has occurred. If our measurement process *f* at time t_1 and t_2 is consistent, we should find a reasonable assessment of the entropy damage that has occurred between these measurement times to be:

$$\Delta S_{f,damage}(t_2, t_1) = \Delta S_f(t_2) - \Delta S_f(t_1) \geq 0 \tag{2.14}$$

where it equals zero if no device degradation is measurable. If we do generate some entropy damage during our measurement process ($t_i + \Delta t_i$), it must be minimal compared to what is generated during the actual aging process between time t_1 and t_2. Then, our entropy measurement difference should be a good indication of the device aging/damage that is occurring between measurement times t_1 and t_2. The actual aging process to the system between time t_1 and t_2, might be a high level of stress applied to the system. Such stress need not be quasistatic. However, the stress must be limited to within reason so that we can repeat our measurement in a consistent manner at time t_2. That is, the stress should not be so harsh that it will affect the consistency of the measurement process *f*.

One might for example have a device aging in an oven in a reliability test, then remove it and make a quasistatic entropy measurement f at time t_1 and then put the device back in the oven later to do another measurement at time t_2. Any resulting measurement difference is then attributed to entropy damage.

(Note that Equation (2.14) is a different statement from entropy flow which many books describe. Here we are concerned with entropy damage over a repeatable measurement process.)

2.3 Intermediate Thermodynamic Aging States and Sampling

In the above, our aging measurements were taken at an initial and final measurement time. These were quasistatic intermediate measurement states where little if any aging may have occurred. We are sampling the aging process. If the system had failed, it would have been in equilibrium with its environment. We did not track how the entropy of the aging process occurred over time. We were only concerned with whether or not degradation occurred. Our detection of the aging process is only limited by our ability to come up with a quasistatic measurement process. In theory, we can even detect if a complex rocket ship is degrading. Each measurement process takes place at a non-aging state at a key tracking point such as an initial, intermediate, or final measurement time. The measurement itself is taken in a small enough window of delta time to observe the state of the system but not cause any significant aging during our observation times.

Given a system, what quasistatic measurement process f will best detect degradation at an intermediate aging state? Part of the problem will always be resolution of our measurement process. Although we now have a tool that can possibly detect aging of a large system, can we make a measurement with enough resolution to observe its degradation? Can we take a partial sample in some way? For example, do we need to completely isolate the system and its environment in the entirety to make a measurement? We are now in a position which challenges our imagination.

The principals will therefore always be valid, but we may be limited by the practicality of the measurement. In the above case, the aging was associated with heat. Not all aging occurs in a manner that allows us to make degradation measurements in this way.

2.4 Measures for System-Level Entropy Damage

We next ask what state variables can be measured as an indicator at the system level for the entropy of aging. In this section we will explore the state system variables of temperature, and in Sections 2.5–2.9 system noise and system failure rate.

2.4.1 Measuring System Entropy Damage with Temperature

The most popular continuous intensive thermodynamic variable at the system level is temperature. In both mechanical and electrical systems, system internal temperature can be a key signature of disorder and increasing entropy. Simply put, if entropy damage increases, typically this is the result of some resistive or frictional heating process. Some examples will aid in our understanding of how such measurements can be accomplished with the aid of temperature observations.

2.4.2 Example 2.3: Resistor Aging

Resistor aging is a fundamental example, since resistance generates entropy. A resistor with value R ohms is subjected to environmental stress over time at temperature T_1 while a current I_1 passes through it for 1 month. Determine a measurement process to find the entropy at two thermodynamic non-aging states before the stress is applied and after it has been applied at times t_{initial} and t_{final}, respectively. Determine if the resistor has aged from the measurement process and the final value for the resistor.

Before and after aging at temperature T_1, we establish a quasistatic measurement process f to determine the entropy change of the resistor at an initial time t_i and final time t_f. A simple method would be to thermally insulate the resistor and pass a current through it at room temperature T_2 for a small time t, and monitor the temperature rise T_3. The internal energy of the resistor with work done on it is:

$$U = W + Q = mCpT_3 = I^2Rt + mCpT_2. \tag{2.15}$$

This yields

$$I^2Rt = mCp(T_3 - T_2) \tag{2.16}$$

and

$$R(t_i) = mCp(T_3 - T_2)/I^2t. \tag{2.17}$$

The entropy for this quasistatic measurement process at initial time t_1 over the time period Δt is considered reversible so that the entropy can be written over the integral:

$$S_f(t_1 + \Delta t) - S_f(t_1) = \int \frac{\delta Q}{T}. \tag{2.18}$$

It is important to note that the entropy change is totally a function of the measurement process f, often termed path dependent in thermodynamics, which is why the notation δQ is used in the integral. The integral for this process ($d(\text{Volume}) = 0$) is:

$$\Delta S_f(t_1) = \int_{T_2}^{T_3} \frac{mC_p(T)dT}{T} = mC_{p,\text{avg}} \ln \frac{T_3}{T_2} \tag{2.19}$$

where T_3 is the temperature rise of the resistor observed after time period Δt.

A month later we repeat this exact quasistatic measurement process at time t_f and find that aging occurred as the temperature observed is now T_4, where $T_4 > T_3$ and the entropy damage change observed over the final measurement time is

$$\Delta S_f(t_2) = mC_{p,\text{avg}} \ln \frac{T_4}{T_2} \tag{2.20}$$

and

$$R(t_f) = mCp(T_4 - T_2)/I^2 t. \tag{2.21}$$

Therefore the entropy damage change related to resistor degradation is

$$\Delta S_{\text{damage}}(t_2, t_1) = mC_{p,\text{avg}} \ln\frac{T_4}{T_3} \tag{2.22}$$

or, in terms of an aging ratio,

$$A_{\text{aging-ratio}} = \ln(T_3/T_2)/\ln(T_4/T_2). \tag{2.23}$$

The resistance change is

$$R(t_f) = \left(\frac{T_4 - T_2}{T_3 - T_2}\right) R(t_i). \tag{2.24}$$

We note

$$R = \rho L/A. \tag{2.25}$$

An analysis of the problem indicates that the material properties and the dimensional aspects need to be optimized to reduce the observed aging/damage that occurred.

One might now ask, so what! We can easily measure the resistance change of the aging process with an ohm meter. On the other hand, we note that our aging measurement was independent of the resistor itself. If we are only looking at aging ratios, we do not even have to know any of its properties. In some cases, we are unable to make a direct measurement; we can detect aging using thermodynamic principles. As well, the next example may be helpful to understand some advantages to this approach for complex systems.

2.4.3 *Example 2.4: Complex Resistor Bank*

In the above example we dealt with a simple resistor, but what if we had a complex resistor bank such as a resistance bridge or some complex arrangement? The system's aging can still be detected even though we may not easily be able to make direct component measurements to see which resistor or resistors have aged over environmental stress conditions. In fact, components are often sealed, so they are not even accessible for direct measurement. In this case, according to our thermodynamic damage theory, if we isolate the system and its environment the total entropy change is, from Equation (2.1):

$$\Delta S_{\text{total}} = \sum_{i=1}^{N} \Delta S_i. \tag{2.26}$$

If we are able to use the exact same measurement process *f* as we did for Section 2.4.2 then for a complex bank the aging ratio is still:

$$A_{\text{aging-ratio}} = \ln(T_3/T_2)/\ln(T_4/T_2). \tag{2.27}$$

2.4.4 System Entropy Damage with Temperature Observations

In thermodynamic terms, if a part is incompressible (constant volume), and if a part is heating up over time t_2 compared to its initial time at t_1 due to some degrading process, then its entropy damage change is given by (e.g., see Equation (2.19)):

$$\Delta S(t_2, t_1) = mC_{\text{avg}} \ln\frac{T_2}{T_1} \tag{2.28}$$

where C_{avg} is the specific heat ($C_p = C_v = C$) and for simplicity we are using an average value.

Then for a system made up of similar type parts, in accordance with Equation (2.28) the total entropy damage is

$$\Delta S_{\text{total}} = \sum_{i=1}^{N} \Delta S_i = \ln\frac{T_2}{T_1} \sum_{i=1}^{N} m_i C_{\text{avg},\,i}. \tag{2.29}$$

This is an interesting result. It helps us to generalize how the state variable of temperature can be used to measure generated entropy damage for a system that is undergoing some sort of damage related to thermal issues. Note that we do not need to know the temperature of each part; temperature is an intensive thermodynamic variable. It is a roughly uniform indicator for the system. Thus, monitoring temperature change of a large system is one key variable that is a major degradation concern for any thermal-type system.

2.4.5 Example 2.5: Temperature Aging of an Operating System

Prior to a system being subjected to a harsh environment, we make an initial measurement M_1^+ at ambient temperature T_1 and find its operation temperature is T_2. Then the system is subjected to an unknown harsh environment. We then return the system to the lab and make a measurement M_2^+ in the exact same way that M_1 was made, where we note it is found to have a different operational temperature T_3. Find the aging ratio where

- $M_1 : T_1 = 300\,\text{K}, T_2 = 360\,\text{K},$
- $M_2 : T_1 = 300\,\text{K}, T_3 = 400\,\text{K}$

$$A_{\text{aging-ratio}} = \ln(400/300)/\ln(360/300) = 1.58.$$

The system has aged by a factor of 1.58. We next need to determine what aging ratio is critical.

2.4.6 Comment on High-Temperature Aging for Operating and Non-Operating Systems

Reliability engineers age products all the time using heat. If this is the type of failure mechanism of concern, then they will often test products by exposing their system to high temperature in an oven. This of course is just one type of reliability test. The product may be operational or non-operational. Often the operational product will add more heat, such as a semiconductor test having a junction temperature rise. Then the true temperature of the semiconductor junction is the junction rise plus the oven temperature. You might ask is there a threshold temperature for aging. The answer is, not really. Temperature aging mechanisms are thermally activated. In a sense, defects are created by surmounting a free energy barrier. Thus, there is a certain probability for a defect to be activated at any given temperature. The most common type of thermally activated aging law is called the Arrhenius Law. The barrier height model is discussed in Chapter 6. However, the Arrhenius Law is found with numerous examples in Chapters 4–10.

2.5 Measuring Randomness due to System Entropy Damage with Mesoscopic Noise Analysis in an Operating System

One might suspect that disorder in the system can introduce some sort of randomness in an operating process of the system. We provide a proof of this in Section 2.6. In this section we assume this randomness is present so we can provide some initial examples first to motivate the reader by demonstrating the importance of noise measurements to assess degradation. Then the reader can first see how it is used to assess aging in operating systems and why it is important. A key example will also be in Section 2.7 on the human heart. First we have to develop some terminology on the measurement process. To start, we define an *operating* system as follows.

An operating system has a thermodynamic process that generates energy (battery, engine, electric motor…), or has energy passing through the system (transistor, resistor, capacitor, radio, amplifier, human heart, cyclic fatigue…).

Compare this to a non-operating aging system such as metal corrosion, ESD damage, sudden impact failure, internal diffusion, or intermetallic growth. In an operating system there is likely a thermodynamic random variable that is easily accessible, such as current, voltage, blood flow, and so forth.

For example, imagine an electronic signal passing through an electrical system of interest that is becoming increasingly disordered due to aging. We might think the disorder is responsible for a signal to noise problem that is occurring. In fact, we can think of the noise as another continuous extensive variable at the system level (scaling with the size of the system damage). Random processes are sometimes characterized by noise (such as signal to noise). Measuring noise is not initially as obvious as say measuring temperature, which is a well-known thermodynamic state variable. In fact, one does not find noise listed in books on thermodynamics as a state variable. Nevertheless, we can characterize one key state of a system using noise.

Entropy Noise Principle: In an aging system, we suspect that disorder in the system can introduce some sort of randomness in an operating system process leading to system noise increase. We suspect this is a sign of disorder and increasing entropy. Simply put, if entropy damage increases, so should the system noise in the operating process.

This is also true for a non-operating system, except that noise is hard to measure so in that case we will focus on operating systems. For example, an electrical fan blade may become wobbly over time. The increase in how wobbly it is can be thought of as noise, not in the acoustic sense. Its degree of how wobbly it is provides a measure of its increasing "noise level." Noise is a continuous random variable of some sort.

Measuring randomness due to aging is sometime easy and sometime subtle. For example, a power amplifier may lose power. This is a simple sign of aging and does not require a subtle measurement technique. However, in many cases aging is not so obvious and hard to detect by simple methods. We would like to have a method for refined prognostics to understand what is occurring in the system before gross observation occur, by which time it may be too late to address impending failure. We therefore define what we will term here (for lack of a better phrase):

> *mesoscopic* noise entropy measurement, any measurement method that can be used to observe subtle system-level entropy damage changes that generate noise in the operating system.

To give an indication of what we mean, we note that a *macroscopic* system contains a large number N of "physical particles," $N > 10^{20}$, say. A *microscopic* system contains a much smaller number N of physical particles, $N < 10^4$, say. For the orders of magnitude between these values of N, a new word has been invented. A *mesoscopic* system is much larger than a microscopic system but much smaller than a macroscopic system. A microcircuit on a chip and a biological cell are two examples of possible mesoscopic systems. Water contained in a bottle is a possible macroscopic system of N molecules of H_2O. A single atomic nucleus might be said to form a microscopic system of N bound neutrons and protons. The point is that we are trying to measure details of a system that are not really on the microscopic or the macroscopic level. Therefore, for lack of a better term, we anticipate that such measurements are more on the mesoscopic level. We do not wish to get too caught up in semantics; the reader should just take it as a terminology that we have invented here to refer to the business of making subtle measurements of entropy damage in a system.

One can also use this term for other *mesoscopic* entropy measurements that are of a different type, for example, scanning electron microscopic (SEM) measurements that are often used to look at subtle damage. This would be another example. However, SEM-type measurements are not of interest in this chapter.

To do this we can treat entropy of a continuous variable using the statistical definition of entropy in thermodynamics employing the concept of differential entropy.

Up until this point we have been using the thermodynamic definition of entropy provided by the experimental definition of entropy. However, we now would like to introduce the statistical definition of entropy. In the simplest form, the statistical definition of entropy is

$$S = -k_B \sum_i P_i \log P_i, \quad \sum_i P_i = 1_i \tag{2.30}$$

where k_B is the Boltzmann constant, P_i is the probability that the system is in the ith microstate, and the summation is over all possible microstates of the system. The statistical entropy has been shown by Boltzmann to be equivalent to the thermodynamic entropy.

If $P_i = 1/N$ for N different microstates ($i = 1,2,\ldots,N$) then

$$S = -k_B \sum_i (1/N)\log(1/N) = k \ln N, \tag{2.31}$$

which increases as N increases. This helps our notion of entropy. The larger the number of N microstates that can be occupied, the greater is the disorder.

However, since we will be dealing now with randomness variables, in a similar manner we need to work with the statistical definitions of entropy for discrete and continuous variable X. In the same sense entropy is a measure of unpredictability of information content. For a discrete and continuous state variable X, the entropy functions in information theory [3, 4] are written

$$\text{Discrete}\, X, s(x): \quad S(X) = -\sum_i f(x_i)\log_2 f(x_i) \tag{2.32}$$

and for the discrete case extended to the continuum

$$\text{Continuous} X, f(x): S(X) = -\int f(x)\log_2[f(x)]dx = -E[\log f(x)] \tag{2.33}$$

where X is a discrete random variable in the discrete case and a continuous random variable in the continuous case, and f is the probability density function. The above equation is referred to as the differential entropy [3].

Note that in differential entropy, the variables are usually dimensionless. So if X = voltage, the solutions would be in terms of $X = V/V_{\text{ref}}$, a dimensionless variable.

Here we are concerned with the continuous variable X having probability density function $f(x)$.

2.5.1 *Example 2.6: Gaussian Noise Vibration Damage*

As an example, consider Gaussian white noise which is one common case of randomness and often reflects many real world situations (note that not all white noise is Gaussian). When we find that a system has Gaussian white noise, the probability density function (PDF) function $f(x)$ is

$$f(x) = \frac{1}{\sqrt{2\pi\sigma^2}}\exp\left(\frac{-(x-\mu)^2}{2\sigma^2}\right). \tag{2.34}$$

When this function is inserted into the differential entropy equation, the result is given by [4]

$$S(X) = \frac{1}{2}\log\left[2\pi e\sigma(x)^2\right]. \tag{2.35}$$

This is a very important finding for system noise degradation analysis, especially for complex systems, as we will see. Such measurements can be argued to be more on the macroscopic level. However, their origins are likely due to collective microstates which we might argue are on the *mesoscopic* level.

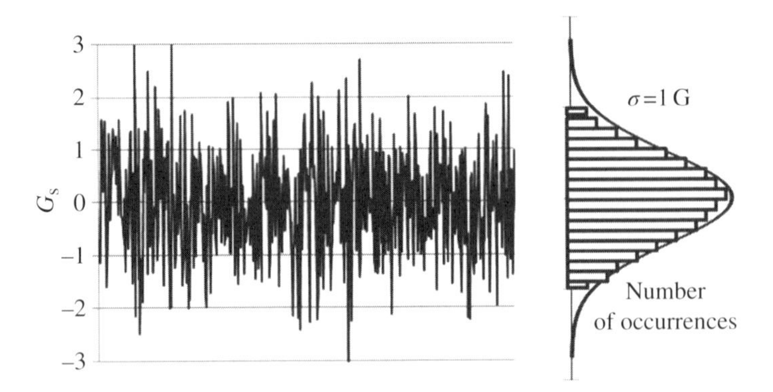

Figure 2.3 Gaussian white noise

To clarify, for a Gaussian noise system the deferential entropy is only a function of its variance σ^2 (independent from the mean μ). This is logical in the sense that, according to the definition of the continuous entropy, it is equal to the expectation value of the log (PDF). For white noise we note that the expectation value of the mean is zero as it fluctuates around zero (see Figure 2.3), but the variance is non-vanishing (see Figure 2.3). Note that the result is also very similar if the noise is lognormally distributed. For a system that is degrading by becoming noisier over time, the entropy damage can be measured in a number of ways where the change in the entropy at two different times t_2 and t_1 is then [2]

$$\Delta S_{\text{damage}} = S_{t2}(X) - S_{t1}(X) = \Delta S(t_2, t_1) = \frac{1}{2}\log\left(\frac{\sigma_{t2}^2}{\sigma_{t1}^2}\right). \tag{2.36}$$

2.5.2 Example 2.7: System Vibration Damage Observed with Noise Analysis

Prior to an engine system being subjected to a harsh environment, we make an initial measurement M_1^+ of the engine vibration (fluctuation) profile which appears to be generating a white noise spectrum in the time domain as is shown in Figure 2.3.

Then the system is subjected to an unknown harsh environment. We again return the system to the lab and make a second measurement M_2^+ in the exact same way that M_1 was made. It is observed that the engine is again generating white noise. The noise measurement spectrum results are

M_1: Engine exhibits a constant power spectral density (PSD) characteristic of $3G_{\text{rms}}$ content in the bandwidth from 10 to 500 Hz.

M_2: Engine exhibits a constant PSD characteristic of $5G_{\text{rms}}$ content in the bandwidth from 10 to 500 Hz.

(Note that standard deviation = G_{rms} for white noise with a mean of zero.) The system noise damage ratio is then:

$$\Delta S_{\text{damage-noise-ratio}} = \log\left(5^2\right)/\log\left(3^2\right) = 1.47. \tag{2.37}$$

Note this is the damage aging *ratio* (see Example 2.5) compared to the damage entropy *change* given by Equation (2.36). This level of entropy damage ratio indicates a likely issue. It would of course be prudent to have more measurements of other similar engines to assess what the statistical true outliers are.

Interestingly enough, noise engineers are quite used to measuring noise with the variance statistic. That is, one of the most common measurements of noise is called the Allan Variance [5]. This is a very popular way of measuring noise and is in fact very similar to the Gaussian Variance, given by:

$$\text{Allan Variance:}\quad \sigma_y^2(\tau)=\frac{1}{2(N-1)}\sum_{i=1}^{N-1}\left[\bar{y}(\tau)_{i+1}-\bar{y}(\tau)_i\right]^2 \tag{2.38}$$

where $N \geq 100$ generally and τ = sample time. By comparison, the true variance is

$$\text{True Variance:}\quad \sigma_y^2(\tau)=\frac{1}{N}\sum_{i}^{N}\left[y(\tau)_i-\bar{y}(\tau)\right]^2. \tag{2.39}$$

We note the Allan Variance commonly used to measure noise is a continuous pair measurement of the population of noise values, while the True Variance is non pair measurement over the entire population. The Allan Variance is often used because it is a general measure of noise and is not necessarily restricted to Gaussian-type noise.

The key results here are that entropy of aging for system noise is dependent on the variance which is also a historical way of measuring noise and is likely a good indicator of the entropy damage of many simple and complex systems.

There are a number of historical options of how noise can be best measured. The author has recently [6, 7] proposed an emerging technology for reliability detection using noise analysis for an operating system. Concepts will be discussed in Section 2.9.

2.6 How System Entropy Damage Leads to Random Processes

Up to this point we have more or less argued that randomness is present in system processes due to an increase in disorder in the system's materials. We now provide a more formal but simple proof of this.

System Entropy Random Process Postulate: Entropy increase in a system due to aging causes an increase in its disorder which introduces randomness to thermodynamic processes between the system and the environment.

Proof: During an irreversible process we can write the time-dependent change in entropy, where it relates to system entropy change, as:

$$\frac{dS_{\text{subsystem}}}{dt}=\sum_{i=1}^{M}\frac{dS_{\text{non-damage-cells},i}}{dt}+\sum_{i=1}^{M}\frac{dS_{\text{damage-cells},i}}{dt}. \tag{2.40}$$

When the entropy is maximum, aging stops and at that point $dS_{\text{subsystem}}(t_i) \sim 0$, then

$$dS_{\text{damage}}(t_i) \approx -dS_{\text{non-damage}}(t_i). \tag{2.41}$$

That is, any non-damage entropy flow to the environment relates to the entropy damage by the above equation. Furthermore, the total cumulative disorder that has occurred in the material has also had non-damage entropy flow to the environment:

$$\sum_{i=1}^{M} dS_{\text{non-damage-cells},\,i} \approx -\sum_{i=1}^{M} dS_{\text{damage-cells},\,i}. \tag{2.42}$$

The disorder in the material is then related to the energy flow to the environment. Furthermore, the damage changes to the system's material is a factor for energy flow. That is, any energy flowing in and out of the system (such as electrical current) can be perturbed by the new disorder in the material. The disorder in energy flow is recognized as a type of randomness and is termed "disordered energy." It is a random process sometimes referred to as system noise as we have described in this chapter.

Next we need to describe what thermodynamic processes are subject to randomness over time.

System Entropy Damage and Randomness in Thermodynamic State Variables Postulate: Entropy damage in the system likely introduces some measure of randomness in its state variables that are measurable when energy is exchanged between the system and the environment.

Proof: We provide here a sort of obvious proof as it helps to clarify the measurement task at hand. Consider a thermodynamic state variable $B(S)$ that is a function of the entropy S, then to a first-order Maclaurin series expansion we have at time t:

$$B(S(t)) = B(0) + S(t)B'(0) + \ldots \tag{2.43}$$

We now consider this at a time later $t+\tau$ when disorder has increased due to aging in the system:

$$B(S(t+\tau) = B(0) + S(t+\tau)B'(0) + \ldots \tag{2.44}$$

Formally, for the thermodynamic variable B we must have

$$B(S(t)) \neq B(S(t+\tau)) \text{ since } S(t+\tau) > S(t). \tag{2.45}$$

However, we also realize that these two are not totally uncorrelated, that is,

$$B(S(t)) \sim B(S(t+\tau)) \text{ for small } \tau \text{ since } S(t) \sim S(t+\tau).$$

Multiplying $B(S(t))$ and $B(S(t+\tau))$ together and taking their expectation value, we have:

$$E[B(S(t))B(S(t+\tau))] = E\left[B(0)^2 + S(t)S(t+\tau)B'(0)^2 + \ldots\right] \tag{2.46}$$

If we set $B(0) = 0$ above and $B'(0) = K$ a constant, and ignore the higher-ordered terms, then the measurement task at hand is

$$E[B(S(t))B(S(t+\tau))] = K^2 E[S(t)S(t+\tau)]. \tag{2.47}$$

This problem reduces to finding what we term as the *Noise Entropy Autocorrelation Function*:

$$R_{S,S}(\tau) = E[S(t),S(t+\tau)] \text{ or } R_{B,B}(\tau) = E[B(t),B(t+\tau)] \tag{2.48}$$

where $R_{S,S}(\tau)$ is the entropy noise autocorrelation function and E indicates that we are taking the expectation value. However, we typically measure $R_{B,B}(\tau)$ where it is some thermodynamic state or related function of the entropy.

Here we have assumed that $S(t+\tau) > S(t)$. But what if $S(t+\tau) = S(t)$? When work is done on the system and energy is exchanged, we can still have randomness. For example, when a system has been damaged, the material is non-uniform since some measure of disorder occurred. What if it was a resistor for example and we passed current through it? At time t, the current path taken may differ from some later time $t+\tau$ as the current path chosen is arbitrarily by the now non-uniform material and is likely a different path in the material to avoid damage sites. The current may even scatter off the damage sites. Here again the current randomness can be measured similarly by the noise autocorrelation function [8], this time for current I where

$$R_{I,I}(\tau) = E[I(t),I(t+\tau)]. \tag{2.49}$$

If $E[I(t),I(t)] \geq E[I(t),I(t+\tau)]$ then we know that it is possible the system damage is creating randomness. In detail, $E[I(t), I(t)]$ is the maximum correlated value so any lower correlated value of the current with its delay τ is likely an indication of system damage. How can we be sure, since it is a subtle measurement? One way is to increase the damage further and make new comparisons. Another way is to take an identical measurement on a similar system that has not experienced damage to have what is termed an experimental control to compare. These concepts are further discussed in detail in Section 2.9.

Formally, there are a number of ways to find the expectation value for the noise autocorrelation function [9]. These are

$$\begin{aligned} R_{s,s}(\tau) = E[S(t),S(t+\tau)] &= \lim_{T\to\infty}\left[\frac{1}{2T}\int_{-T}^{T} S(t)S(t+\tau)dt\right] \\ &= \int_{-\infty}^{\infty} S(t)S(t+\tau)dt = \int\int_{0}^{\infty} s_1 s_2 P(s_1,s_2)ds_1 ds_2 \end{aligned} \tag{2.50}$$

where in the last integral $s_1 = S(t)$, $s_2 = S(t+\tau)$, and $P(s_1,s_2) =$ joint PDF of s_1 and s_2. Note that the autocorrelation function can be normalized by dividing it by its maximum value that occurs when $\tau = 0$, that is:

$$R_{s,s}(\tau)_{\text{norm}} = \frac{E[S(t),S(t+\tau)]}{E[S(t),S(t)]} = \frac{\int_{-\infty}^{\infty} S(t)S(t+\tau)dt}{\int_{-\infty}^{\infty} S(t)^2 dt}, \tag{2.51}$$

where

$$-1 \leq R_{s,s}(\tau)_{\text{norm}} \leq 1.$$

Here $E[S(t), S(t)]$ is the maximum value. We see that any thermodynamic variable that is part of the entropy exchange process between the environment and the system is subjected to randomness stemming from system aging disorder, of which one measurement type is to look at its autocorrelation function.

You might be asking at this point in time, well what does this all mean? In simple terms, an example of an ordered process is people walking out the door of a movie theatre at a rate of one person every 5 s. This is a non random process. Then later we note some randomness; we see that people walk out at an average of 5 s intervals, but occasionally there are people walking out at random times of 1, 2, 3, 4, 6, 7 s intervals. This is initially unpredictable. Prior to randomness being introduced, the autocorrelation was $R = 1$. Once randomness occurred, R ranged over $-1 < R < 1$. As more and more randomness is introduced, we expect R to get closer and closer to zero.

Thus, autocorrelation of key state variables is likely a good *mesoscopic* measure of subtle damage: it is a measure of how random the process has become as degradation increases. If the randomness is stabilized, it is called a stationary random process. However, if randomness in the process is increasing over time, it is a non-stationary random process.

2.6.1 Stationary versus Non-Stationary Entropy Process

Measured data can often be what is termed non-stationary, which indicates that the process varies in time, say with its mean and/or variances that change over time. Although a non-stationary behavior can have trends, these are harder to predict than a stationary process which does not vary in time.

Therefore, non-stationary data are typically unpredictable and hard to model for autocorrelated analysis. In order to receive consistent results, the non-stationary data need to be transformed into stationary data. In contrast to the non-stationary process that has a variable variance and a mean that does not remain near, or return to, a long-run mean over time, the stationary process reverts around a constant long-term mean and has a constant variance independent of time.

Although aging processes are non-stationary, they may vary slowly enough so that in the typical measurement time frame they are stationary. Furthermore, disorder is cumulative which means there is a memory in the process. This means there is a likely trend. The issue is: this is a new way of looking at aging. There are, to this author's knowledge, few studies on the subject. We describe below one good study on aging of the human heart to help the reader understand the potential importance of noise autocorrelation measurements.

We summarize below what we would anticipate from such a study.

1. Entropy randomness mesoscopic measurements.
2. If a system has a thermodynamic process that generates energy (battery, engine, electric motor,...) or has energy passing through the system (transistor, resistor, capacitor, radio, amplifier,...), the subtle irreversibilities occurring in the system are likely generating noise in the operating system.
3. Non-damage flow exchange that is part of the exchange process related to entropy damage exhibits randomness that we term noise.

4. For such subtle system-level damage, we employ the method that we term here as *mesoscopic* entropy noise measurement analysis to detect subtle changes in the system due to entropy damage. Such damage is not easily detectable by other means.
5. This randomness, if measured by one or more thermodynamic state variables, is likely observable from the entropy exchange process with the aid of the autocorrelation function of the state variable.
6. Over time, as the system ages, we suspect that the increase in disorder in the system will tend to reduce the autocorrelation result so that the normalized value of $R_{S,S}(\tau)$ will approach 0 when the system is in a state of maximum entropy (disorder).
7. The aging occurring is likely on a longer time scale than the measurement observation time so that the measurement process can be treated as stationary.

2.7 Example 2.8: Human Heart Rate Noise Degradation

Although noise degradation measurements are difficult to find, one helpful example of noise autocorrelation analysis of an aging system found by the author is in an article by Wu *et al.* [10] on the human heart.

Here heart rate variability was studied in young, elderly, and congestive heart failure (CHF) patients. Figure 2.4 shows noise limit measurements of heart rate variability. We note that heart rate noise limit variability between young and elderly patients are not dramatically different compared to what is occurring in patients with CHF. Although this is not the same system (i.e., different people), such measurements can be compared using noise analysis described in this section. This is further illustrated in Figure 2.5 showing noise variability in heartbeats of young subjects compared with CHF patients. This is an example of entropy damage comparison in a complex human heart aging system between a good and a failing system observed well prior to catastrophic failure. This reference [10] shows a variation of how our example in Section 2.5 and damage noise entropy mesoscopic measurements in general can be implemented, and would be helpful as a detection method of a system's thermodynamic degradation state.

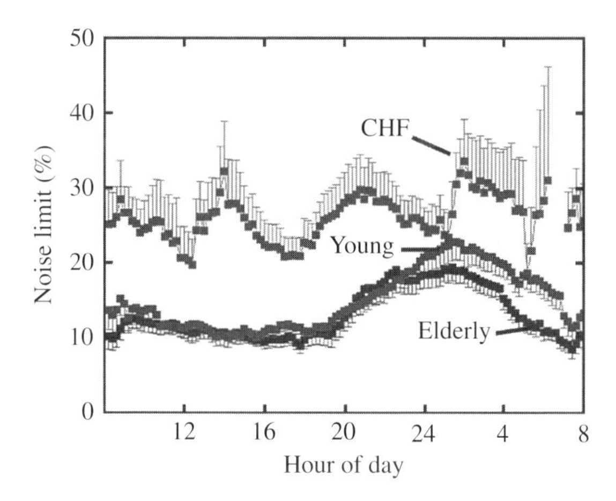

Figure 2.4 Noise limit heart rate variability measurements of young, elderly, and CHF patients [10]

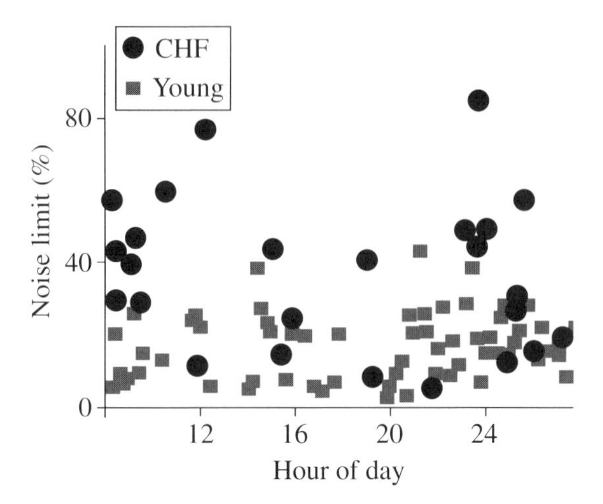

Figure 2.5 Noise limit heart rate variability chaos measurements of young and CHF patients [10]

2.8 Entropy Damage Noise Assessment Using Autocorrelation and the Power Spectral Density

In this section we would like to overview more information about the autocorrelation function and detail its Fourier transform to the frequency domain. This is because there is a lot of related work presented in the frequency domain, which is not framed in the category of degradation, which we anticipate is related in many ways. We start by restating the autocorrelation function here for convenience

$$R_{yy}(\tau) = \int_{-\infty}^{\infty} y(t)y(t+\tau)\,dt \tag{2.52}$$

which was originally described in Equation (2.50). Some properties of the autocorrelation function are:

$$R(0) = \text{Variance} = \sigma^2 = \text{RMS}^2 \geq 0 \tag{2.53}$$
$$R(\tau) = R(-\tau)$$
$$R(0) \geq |R(\tau)|$$
$$-1 \leq \frac{R(\tau)}{R(0)} \leq 1.$$

A graphical representation of the autocorrelation function is shown in Figure 2.6.

The Fourier transform of the autocorrelation function provides what is termed the *PSD Spectrum*, given by

$$\text{PSD}(f) = S_{yy}(f) = \int_{-\infty}^{\infty} R_{yy}(\tau)e^{-j2\pi f\tau}d\tau, \tag{2.54}$$

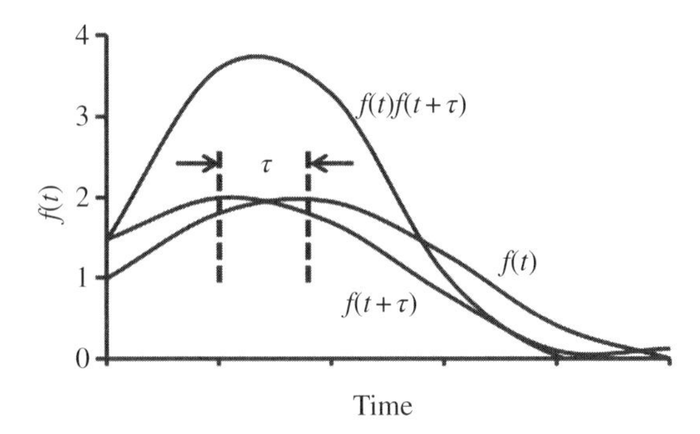

Figure 2.6 Graphical representation of the autocorrelation function

Table 2.1 Common time series transforms

Time series	Autocorrelation	PSD
Pure sine tone	Cosine	Delta functions
Gaussian	Gaussian	Gaussian
Exponential	Exponential decay	Lorentzian $1/f^2$
White noise	Decay	Constant
Delta function	Delta function	Constant

where f is the frequency. Noise measurements are often easily made with the aid of a noise measurement system that can be purchased. These systems do the math for you, which uncomplicates things. There are some common transforms that are worth noting; these are listed in Table 2.1.

For a well-behaved stationary random process, the power spectrum can be obtained by the Fourier transform of the autocorrelation function. This is called the Wiener–Khintchine theorem. Generally, knowledge of one (either R or S) allows the calculation of the other. However, some information can be lost in conversion between say $S(\omega)$ to $R(\tau)$. As a simple example of a transform and some issues, consider Figure 2.7 that illustrates a Fourier transform of a sine wave with some randomness in the frequency variation. Because of the randomness, this is not a pure sine tone.

We note that it is difficult to see the variability in frequency in the time series of Figure 2.7a. However, the variability is easily observed from the width of the frequency spectrum in Figure 2.7b. If the signal was a pure sine tone, the Fourier transform would be a delta-function (spike) -like form. Second we note that knowledge of the frequency spectrum will not indicate which wave occurred first; that is clearly indicated in the time series spectrum where the 100 Hz wave occurred first. Therefore, whenever doing this type of analysis, it can be important to look at both the time series and frequency spectrum.

2.8.1 Noise Measurements Rules of Thumb for the PSD and R

Since people are not familiar with noise measurements, we provide some simple rules of thumb to remember.

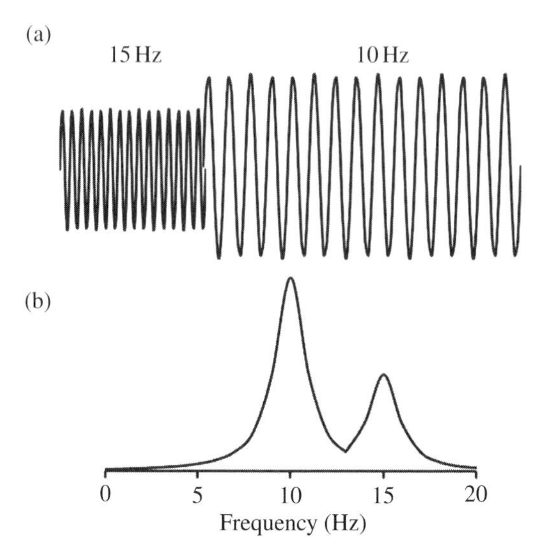

Figure 2.7 (a) Sine waves at 10 and 15 Hz with some randomness in frequency; and (b) Fourier transform spectrum. In (b) we cannot transform back without knowledge of which sine tone occurred first

- As the absolute value of the autocorrelation function decreases to zero, $|R_{\text{norm}}| \to 0$, noise levels are increasing in the system.
- As the PSD(f) levels increase, noise is increasing in the system.
- The slope of the autocorrelation function is related to the slope of the PSD spectrum. As the autocorrelation slope increases, the PSD slope decreases. The slope is an indication of the type of system-level noise characteristic detailed in the next section.

2.8.2 *Literature Review of Traditional Noise Measurement*

There are countless noise measurements in the literature [11]. Here we would like to review well-known noise measurements as we feel it will be helpful in eventually working with degradation problems.

We believe that noise analysis can be crucial in prognostics as we suspect that noise analysis will be a highly important contributing tool. The problem is unfortunately that, at the time of writing of this book, there seems to be little work in the area of noise degradation analysis similar to the human heart study that we have illustrated (i.e., looking at noise as a system ages).

It is clear from our example on the human heart and as we have theoretically shown in Section 2.5 that the noise variance increases with system damage as it is a dependent variable of entropy. This is an indication that the autocorrelation function for system noise is becoming less and less correlated over time for non-stationary processes. As such, we anticipate that the frequency spectrum might have a tendency to higher levels of noise values.

Put another way, as the autocorrelation function becomes less correlated, the PSD spectrum increases. The slope of the PSD curve is an indication of the type of system-level noise observed. Note the slopes of the autocorrelation function and the PSD curve are related. We can clearly see that as the autocorrelation slope decreases (from Figures 2.8b, 2.9b, and

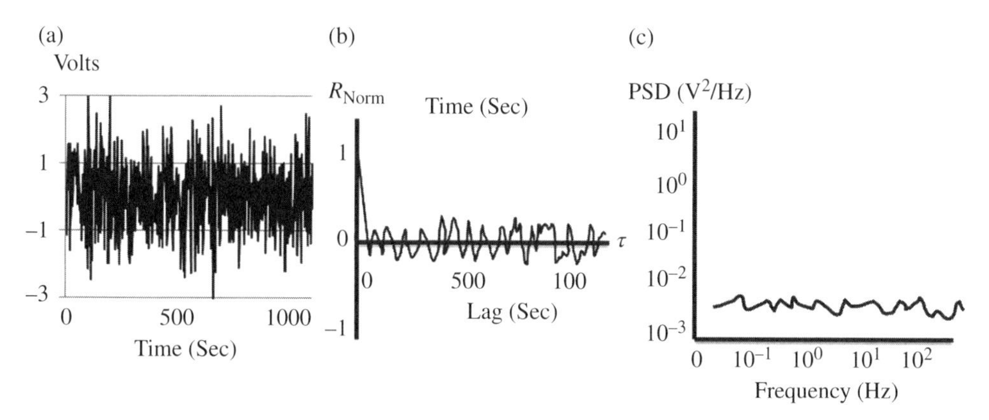

Figure 2.8 (a) White noise time series; (b) normalized autocorrelation function of white noise; and (c) PSD spectrum of white noise

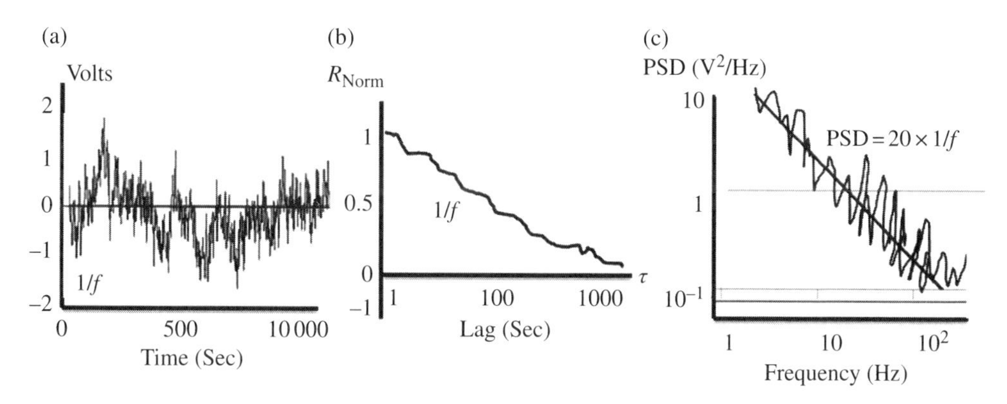

Figure 2.9 (a) Flicker (pink) $1/f$ noise; (b) normalized autocorrelation function of $1/f$ noise; and (c) PSD spectrum of $1/f$ noise

2.10b), the slope of the PSD curve increases (from Figures 2.8c, 2.9c, and 2.10c). Note in Figure 2.8b the slope is hard to see as it decreases quickly near zero time lag. It is possible that the slope may also change in a non-stationary process; we at least expect the noise level to increase. The problem is, as we have mentioned, there is little work in this area to date for non-stationary noise processes.

Noise is sometimes characterized with color [11]; the three main types in this category are white, pink, and brown noise. Each has a specific type of broadband PSD characteristic. We will first overview these then we will describe some reliability-related type of noise measurements. Each type has an interesting sound that is audible in the human audio frequency range (~20 Hz to 20 kHz) of the spectrum.

Figure 2.8 is a well-known time series that we previously discussed for white noise; see Section 2.5.1 and Figure 2.3. In Figure 2.8b we see the autocorrelation function that starts at 1 ($\tau = 0$) and quickly diminishes, becoming uncorrelated. The PSD spectrum appears to be flat (constant) over the frequency region. However, the variance is non-vanishing. Much work has been done in the area of thermal noise, also known as Johnson noise or Johnson–Nyquist noise, which has a white noise spectrum (see Section 2.9.1).

Figure 2.9 is a well-known phenomenon called $1/f$ noise, also referred to as pink noise or flicker noise. While many measurements have been carried out and many theories have been proposed, no one theory uniquely describes the physics behind $1/f$ noise. It is generally not well understood, but is considered by many to be a non-stationary process [11, 12]. From the time series, the autocorrelation function has more correlation over the lag time τ compared to white noise. The fact that it is called $1/f$ has to do with the slope of the PSD line as shown in Figure 2.9c. In the PSD plot, we see that there is more "power" at low frequencies and appears to diverge at 0.

If we compare figures again, we see that a rapid variation in the time series such as White noise in Figure 2.8 is harder to correlate than a slower variation such as Figure 2.10. In terms of the PSD spectrum, we carefully plotted $1/f$ and $1/f^2$ in Figures 2.9 and 2.10c so one can see the relative slope change on the same scale. This is also compared in Figure 2.11. The slope yielding $1/f$ has been measured over longer periods of lag of just 1 s. There are reports of lags say over 3 weeks with a frequency measured down to $10^{-6.3}$ Hz, and the slope continuous to diverge as $1/f$ with no change in shape [13]. Using geological techniques, the slope of $1/f$ has been computed down to 10^{-10} Hz or 1 cycle in 300 years [14].

The author anticipates that, if the measurement was taken on semiconductors that had been subjected to parametric aging, the following would hold.

> A system undergoing parametric degradation will have its noise level increase. Studies suggest (see Section 2.8.3) for example that
>
> $$\text{Relative PSD noise level} \sim \text{PSD}_V(t)^\gamma$$
>
> where t is time, subscript V is a system-level state parameter (such as voltage; see Table 1.1), and γ is a power related to the aging mechanism. The system noise would become more uncorrelated yielding a decrease in R for the autocorrelation function and a higher noise level in the PSD spectrum (see Section 2.8.3).

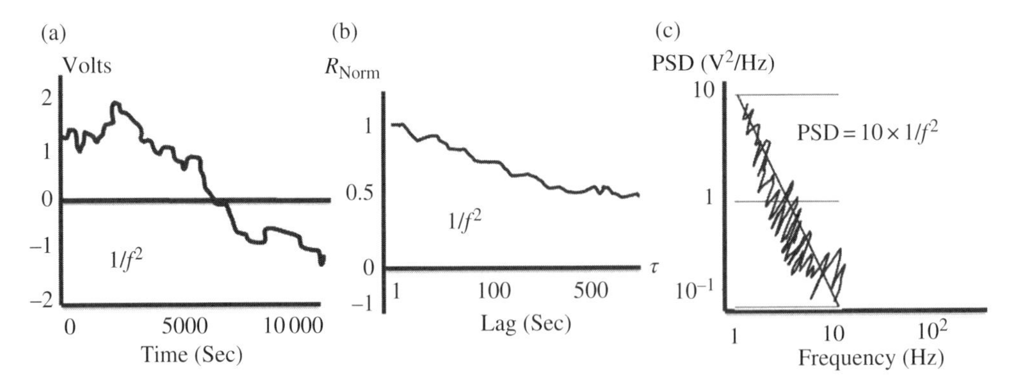

Figure 2.10 (a) Brown $1/f^2$ noise; (b) normalized autocorrelation function of $1/f^2$ noise; and (c) PSD spectrum of $1/f^2$ noise

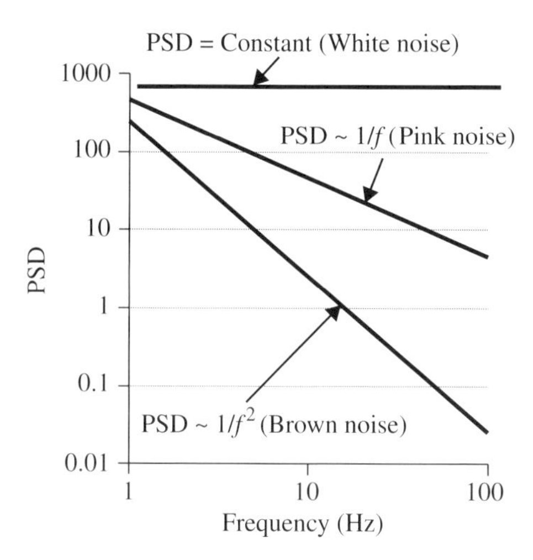

Figure 2.11 Some key types of white, pink, and brown noise that might be observed from a system

This of course is a statement that the time series changes in aging and, at maximum entropy where failure occurs, we would observe the maximum noise characteristic. There are a number of reasons for this. One key reason is that the material degrades and in effect its geometry is changing due to increasing disorder. For example, a resistor will effectively have less area for current to pass through as the material degrades. In effect the resistor is becoming smaller. We will see in Section 2.8.3 when we look at resistor $1/f$ type noise characteristics published in the literature that resistors with less area display larger noise levels. Resistors with smaller area handling the same current will have current crowding, which increases noise.

The interesting issue with $1/f$ noise [11, 12] is that it has been observed in many systems:

- the voltage of currents of vacuum tubes; Zener diodes; and transistors;
- the resistance of carbon resistors; thick-film resistors; carbon microphones; semiconductors; metallic thin-films; and aqueous ionic solutions;
- the frequency of quartz crystal oscillators; average seasonal temperature;
- annual amount of rainfall;
- rate of traffic flow;
- rain drops;
- blood flow;
- and many other phenomena.

Figure 2.10 is another well-known phenomenon called $1/f^2$ noise, also referred to as brown (sometimes red) noise, and is a kind of signal-to-noise issue related to Brownian motion. This is sometimes referred to as random walk which is displayed in Figure 2.10a. Here again the slope of the PSD spectrum goes as $1/f^2$ (hence the name).

Figure 2.11 compares noise-type slopes on the same graph scale so one can see the relative slopes for a PSD plot.

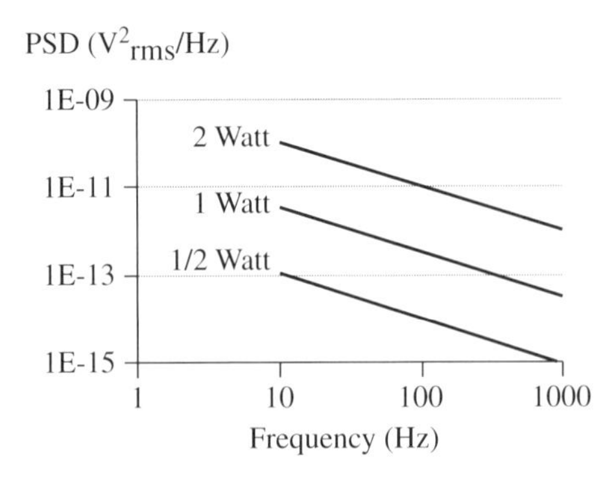

Figure 2.12 1/*f* noise simulations for resistor noise. Note the lower noise for larger resistors (power of 2) and higher noise for smaller resistors (power of 1.5)

2.8.3 *Literature Review for Resistor Noise*

We have mentioned in Section 1.7 that entropy can be produced from resistance. Resistance is a type of friction which increase temperature, creates heat, and often entropy damage. In the absence of friction, the change in entropy is zero. Therefore, it makes sense in our discussion on noise as an entropy indicator that we look at noise in electrical resistors. One excellent study is that of Barry and Errede [15]. These authors found for 1/*f* noise in carbon resistors that noise measurements on 0.5, 1, and 2 W resistors indicated that the 0.5 W resistors had the largest noise and the 2 W resistors had the least noise. Figure 2.12 illustrates the PSD noise level for carbon resistors as well as the 1/*f* noise type slopes for these resistors.

We see that resistors with reduced volume or cross-sectional area produce more noise. Other authors have found that 1/*f* noise intensity for thick-film resistors varies directly with resistance [16, 17]. This is a similar observation as resistor wattage goes as its volume. We can interpret this as an increase in resistance produces higher levels of entropy, increasing disorder, and increasing system-level noise. From an electrical point of view, smaller resistors can produce higher current densities increasing noise. The fact that the noise characteristics are 1/*f*-type (pink noise) is related to the type of noise phenomenon in the system.

2.9 Noise Detection Measurement System

As an example of how we can use the autocorrelation method on an operating or non-operating system, we can start with the standard electronic system for such measurements. We feel that an active on-board autocorrelation measurement of key telltale parameters can be helpful for prognostics in determining a system's wellness in real time [6, 7]. Such software methods exist for measurements [7]. Figure 2.13 illustrates a system that can be used for an active measurement process using an electronic multiplier and a delayed signal.

Figure 2.13 shows an initial, say engine vibration, sample at time t, $y(t)$ and with delay τ. This would be a stationary measurement process of the system. The two signals are mixed in the multiplier and averaged. The autocorrelation results can be stored as, say, an engine's vibration noise sample. Then later we can take another sample at time t_2 after the system has been

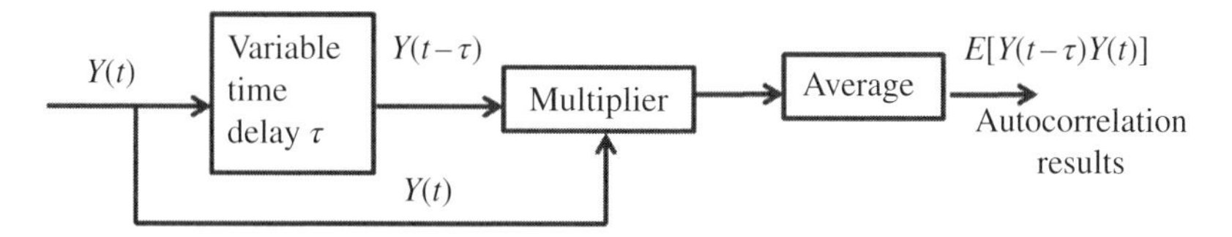

Figure 2.13 Autocorrelation noise measurement detection system

exposed to a stress and compare to see if the noise level has increased. Even though the aging process is non-stationary over time, the actual measurements are stationary as negligible aging occurs during the measurement process. The key difference here is that one is comparing the signals over time. The figure is simplified for a conceptual overview. More sophisticated mixing methods exist [6, 7], as well as prognostic software [7]. The autocorrelated results can of course also be analyzed in the frequency domain with typical PSD magnitude. For random vibration it would have a noise metrics (G_{rms}^2/Hz) with G_{rms} spectral content, for voltage noise V^2/Hz, etc. Here for engine vibration, one would use accelerometers to generate the $y(t)$ voltage signals (where an accelerometer is a device that converts vibration to voltage), and also look at possible resonance assessment with Q values found from the spectra observations (see Section 4.3.5 for an explanation of resonance and Q effect). Assessment of this type is not limited to engine noise but any type of noise. Noise measurements can be made for electronic circuits, engines, fluid flow, human blood flow (as per the study we presented in Section 2.7), etc.

Such noise measurements can also include amplitude modulation (AM), frequency modulation (FM), or phase modulation noise analysis. For example, a sinusoidal carrier wave can be written

$$A(t) = A_{max} \sin(\omega t + \phi) \tag{2.55}$$

where A_{max} is its maximum value, ω is the angular frequency, and ϕ is the phase relation. We see that noise modulation can occur to the amplitude, frequency, or phase.

Once noise is captured, one must use engineering/statistical judgment to assess the threshold of the noise issue that can be tolerated before maintenance is warranted.

Statistically, it is easier to judge when maintenance is needed based on a number of such systems. As we gain experience with the type of system we are measuring, the information we obtain is easier to interpret. Therefore, analysis of numerous units when assessed will help determine normality and when maintenance is needed. Typically, a good sample size is likely 30 or greater as the variance of a distribution statistically is known to stabilize for this sample size.

2.9.1 System Noise Temperature

It is not surprising that noise and temperature are related. For example, thermal noise also known as Johnson–Nyquist noise has a white noise spectrum. The noise spectral density in resistors goes directly with temperature and resistance according to the Nyquist theorem, defined as

$$S_{V^2}(f) = 4k_B TR$$

where R is resistance, T is the temperature, and k_B is Boltzmann's constant [11]. Simply put, as temperature and/or resistance increases, so too will the noise level observed in resistors. Thermal noise is flat in frequency as there is no frequency dependence in the noise spectral density; therefore it is a true example of white noise.

2.9.2 Environmental Noise Due to Pollution

We have asserted that temperature, noise, and failure rate are some good key system thermodynamic state variables. However, depending on how you define the system and the environment, they can also be state variables for the environment. An internal combustion (IC) engine is a system which interacts with the environment causing entropy damage to the environment. This pollution damage and that of other systems, on a global scale, are unfortunately measurable as meteorologists track global climate change. The temperature (global warming) effect is the most widely tracked parameter. However, it might also make sense from what we determined above to look at environmental noise degradation. As our environment ages, we might expect from our above finding that its variance will increase. Some indications of environmental variance change are larger swings in wind, rain, and temperature, causing more frequent and intense violent storms. It might be prudent to focus on such environmental noise issues and not just track global temperature rise [2].

2.9.3 Measuring System Entropy Damage using Failure Rate

So far we have two state variables for measuring aging. Another possible state variable, somewhat atypical but often used by engineers, is the reliability metric called the failure rate λ. We do not think of the failure rate as a thermodynamic state variable. In fact, to any reliability engineer, it is second nature that the failure rate is indeed a key degradation unit of measure that helps to characterize a system in terms of its potential for failure over time. For consistency, we would like to put it in the context of entropy damage to see what we find. For complex systems, the most common distribution used by reliability engineers comes from the exponential probability density function:

$$f(x) = \lambda \exp(-\lambda x). \tag{2.56}$$

The differential entropy can be found easily (see Equation (2.33)) from the negative of the expectation of the natural logarithm of $f(x)$, defined [9]:

$$\begin{aligned} S &= -E[\ln(\lambda \exp - \lambda x)] \\ &= -E[\ln(\lambda) + \ln(\exp - \lambda x)] \\ &= -\ln(\lambda) + \lambda E[x] = \ln(1/\lambda) + 1, \end{aligned} \tag{2.57}$$

where for the exponential distribution the mean $E[x] = 1/\lambda$. A way to view this is that the exponential distribution maximizes the differential entropy over all distributions with a given mean and supported on the positive half line $(0,\infty)$. This is not a very exciting result, but does provide some physical insight from an entropy point of view on why it is one of the more popular distribution used in reliability for complex systems. So in general, we turned the problem around to the fact that any PDF will maximize the entropy subjected to certain constraints, in this case $E[x] = 1/\lambda$. For a normal distribution it is subjected to a mean and sigma over the interval $(-\infty,\infty)$. In the case of white noise, the results proved helpful in our noise analysis (see Section 2.5).

For the exponential distribution, we note that if the environmental stress changes we can measure the entropy change. In this case, the entropy damage change would be for this system under two different environments:

$$\Delta S = \ln(\lambda_{E1}/\lambda_{E2}) \tag{2.58}$$

where the subscripts $E1$ and $E2$ are for environments 1 and 2.

2.10 Entropy Maximize Principle: Combined First and Second Law

Typically, entropy change is measured and not absolute values of entropy. We may combine the general first law statement

$$dU = \delta Q + \delta W = \delta Q + \sum_a Y_a dX_a \tag{2.59}$$

with the statement of the thermodynamic second law

$$\delta Q = TdS$$

(for reversible processes, or equilibrium condition). The full statement of the two thermodynamic laws involves *only exact differential forms*. For a general system the combined statement reads, for a reversible process,

$$\begin{aligned} U &= U(S, X_1, \ldots, X_n), \\ dU &= TdS + \sum_a Y_a dX_a. \end{aligned} \tag{2.60}$$

The internal energy U is a function of the entropy and the generalized displacements. During a quasistatic process, the energy change may be decomposed into a term representing the heat flow TdS from the environment into the system plus the work $\sum_a Y_a dX_a$ done by the environment on the system.

In terms of the entropy function, we have

$$\begin{aligned} S &= S(U, X_1, \ldots, X_n), \\ dS &= \left(\frac{1}{T}\right) dU - \sum_a \left(\frac{Y_a}{T}\right) dX_a. \end{aligned} \tag{2.61}$$

The entropy maximum principle in terms of aging: Given a set of physical constraints, a system will be in equilibrium with the environment if and only if the total entropy of the system and environment is at a maximum value:

$$S_{\text{total}} = S_{\max} = S_{\text{system}} + S_{\text{env}}. \tag{2.62}$$

At this point, the system has lost all its useful work, maximum damage has occurred where $dS_{\text{total}} = 0$, and no further aging will occur.

In an irreversible process $TdS > \delta Q$, but $\delta Q = dU + PdV$ for pressure volume work so, for example, $TdS > dU + PdV$.

The following three examples will help to explain the entropy maximum principle.

2.10.1 *Example 2.9: Thermal Equilibrium*

Consider a system that can exchange energy with its environment subject to the constraint of overall energy conservation. The total conserved energy is written

$$U_{\text{total}} = U + U_{\text{env}} \tag{2.63}$$

where we have dropped the "system" subscript. Under a quasistatic exchange of energy we have

$$dU_{\text{total}} = dU + dU_{\text{env}} = 0.$$

Then, from Equation (2.60) we have

$$dS_{\text{total}} = \frac{dU}{T} + \frac{dU_{\text{env}}}{T_{\text{env}}} + \frac{YdX}{T} + \frac{Y_{\text{env}} dX_{\text{env}}}{T_{\text{env}}}. \tag{2.64}$$

Let's apply a physical situation to this equation; we can use the conjugate pressure P and volume V thermodynamic work function PdV for YdX. Figure 2.14 displays this example. In the figure we imagine a perfect insulating cylinder so that heat does not escape. It is divided into two sections by a piston which can move without friction. The piston is also a heat conductor. We take the left side to represent the environment and the right side that of the system.

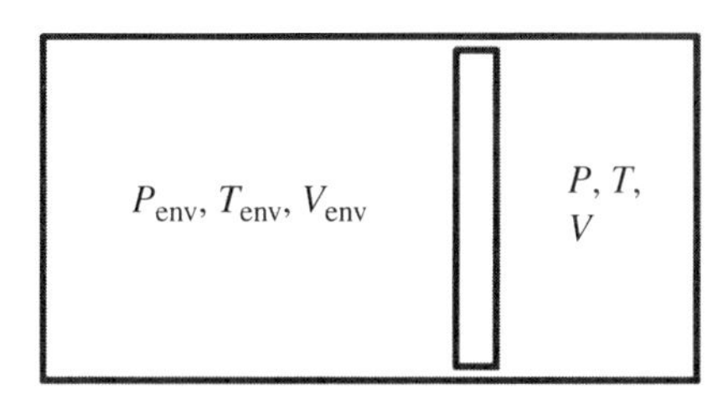

Figure 2.14 Insulating cylinder divided into two sections by a frictionless piston

Since $dU = -dU_{\text{env}}$ and, when the volume changes incrementally, we will also have $dV = -dV_{\text{env}}$, we can write

$$dS_{\text{total}} = \left(\frac{1}{T} - \frac{1}{T_{\text{env}}}\right)dU + \left(\frac{P}{T} - \frac{P_{\text{env}}}{T_{\text{env}}}\right)dV \tag{2.65}$$

and

$$dS_{\text{total}} \geq 0. \tag{2.66}$$

In order to ensure that the total entropy goes to a maximum, we have positive increments under the exchange of energy in which the system energy change is dU. If the environmental temperature is more than the system temperature, that is, $T_{\text{env}} > T$, then this dictates that $dU > 0$ and energy flows out of the environment into the system. If the environmental temperature is less than the system temperature, that is, $T_{\text{env}} < T$, then this dictates that $dU < 0$. The energy then flows out of the system into the environment from the higher-temperature region to the lower-temperature region. A similar argument can be made for the pressure; if $P > P_{\text{env}}$ then $dV > 0$ and the piston is pushed to the left. If the entropy is at the maximum value, then the first order differential vanishes to yield

$$dS_{\text{total}} = \left(\frac{1}{T} - \frac{1}{T_{\text{env}}}\right)dU + \left(\frac{P}{T} - \frac{P_{\text{env}}}{T_{\text{env}}}\right)dV = 0. \tag{2.67}$$

At equilibrium the energy exchange will come to a halt when the environmental temperature is equal to the system temperature, that is, $T = T_{\text{env}}$, and $P = P_{\text{env}}$. This is in accordance with the maximum entropy principle, which occurs when the system and the environment are in equilibrium. This of course is an irreversible process; in thermodynamic terms there is a tendency to maximize the entropy and the pressure and temperature will not seek to go back to the non-equilibrium states. In terms of aging, we might have a transistor that is being cooled by a cold reservoir. Pressure may be related to some sort of mechanical stress. If the transistor fails due to the mechanical stress, the transistor comes to equilibrium with the environment as the pressure is released and the pressure and temperature of the system and environment are equal.

2.10.2 Example 2.10: Equilibrium with Charge Exchange

Shown in Figure 2.15 is a system (capacitor) in contact with an environment (battery).

A (possibly) non-linear capacitor is connected to a battery. The capacitor and the battery can exchange charge and energy subject to conservation of total energy and total charge. Then with these constraints we wish to maximize

$$S_{\text{total}} = S(U, q) + S_{\text{env}}(U_{\text{env}}, q_{\text{env}}) \tag{2.68}$$

with the total energy and charge

$$U_{\text{total}} = U + U_{\text{env}} \tag{2.69}$$

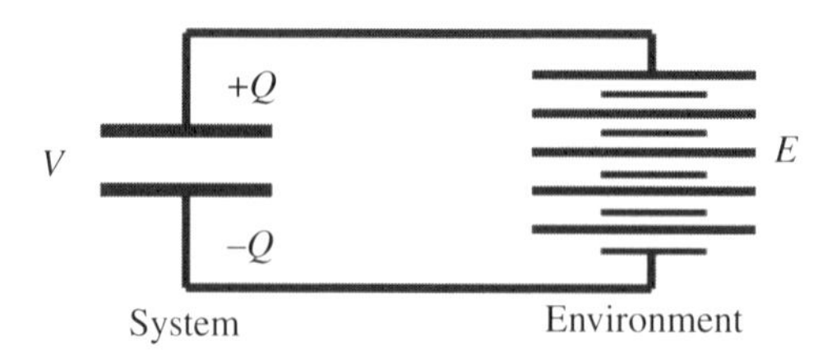

Figure 2.15 System (capacitor) and environment (battery) circuit

and

$$q_{\text{total}} = q + q_{\text{env}}, \tag{2.70}$$

obeying

$$dU_{\text{total}} = dU + dU_{\text{env}} = 0 \tag{2.71}$$

and

$$dq_{\text{total}} = dq + dq_{\text{env}} = 0. \tag{2.72}$$

The entropy for the capacitor and battery requires

$$dS_{\text{total}} = dS + dS_{\text{env}}. \tag{2.73}$$

Then we can write from the combined first and second law equation

$$dS_{\text{total}} = \left(\frac{dU}{T} - \frac{Vdq}{T}\right) + \left(\frac{dU_{\text{env}}}{T_{\text{env}}} - \frac{Edq_{\text{env}}}{T_{\text{env}}}\right) \tag{2.74}$$

or, equivalently,

$$dS_{\text{total}} = \left(\frac{1}{T} - \frac{1}{T_{\text{env}}}\right)dU + \left(\frac{E}{T_{\text{env}}} - \frac{V}{T}\right)dq \tag{2.75}$$

and

$$dS_{\text{total}} \geq 0. \tag{2.76}$$

As before when $T_{\text{env}} > T$, then $dU > 0$ energy flows out of the environment into the capacitor system. For $E > V$, $dq > 0$ and charge is flowing out of E (battery) into V (capacitor system). For equilibrium when the entropy is maximum, $T = T_{\text{env}}$ and $V = E$, so that $dS_{\text{total}} = 0$. The battery has in a sense degraded its energy in charging the capacitor so that, in the absence of capacitor leakage, no more useful work will occur. Again, the entropy is maximized in the irreversible process. There is no tendency for the charge to flow back into the battery.

2.10.3 Example 2.11: Diffusion Equilibrium

In Chapter 8 for spontaneous diffusion processes we make an analogy to Example 2.10. We find for equilibrium of diffusion into an environment that when the entropy is maximum $T = T_{env}$ and $\mu = \mu_{env}$. Here T, T_{env}, μ, and μ_{env} are system and environmental temperatures and chemical potentials, respectively, prior to diffusion. As well the pressure ($P = nRT/V$) is the same in system and environmental regions at equilibrium where the system and environment are shown in Figure 8.1. Therefore, atoms or molecules migrate so as to remove differences in chemical potential. Diffusion ceases at equilibrium, when $\mu = \mu_{env}$. Diffusion occurs from an area of high chemical potential to low chemical potential. The reader may be interested to work the analogy and compare results to that found in Section 8.1.

2.10.4 Example 2.12: Available Work

Consider a system in a state with initial energy U and entropy S which is not in equilibrium with its neighboring environment. We wish to find the maximum amount of useful work that can be obtained from the system if it can react in any possible way with the environment [4]. Let's say the system wants to expand to come to equilibrium with the environment and we have a shaft on it that is doing useful work (see Figure 2.16).

From the first law we have

$$dU = -\delta W - P_0 dV - \delta Q. \tag{2.77}$$

The entropy production is

$$dS_{gen} = dS_{damage} - \frac{\delta Q}{T_0} \tag{2.78}$$

where dS_{damage} is the increase in entropy damage and $\delta Q/T_0$ is the entropy outflow. Combining these two equations then the useful shaft work is

$$\delta W = -dU - P_0 dV - T_0\left(dS_{gen} - dS_{damage}\right). \tag{2.79}$$

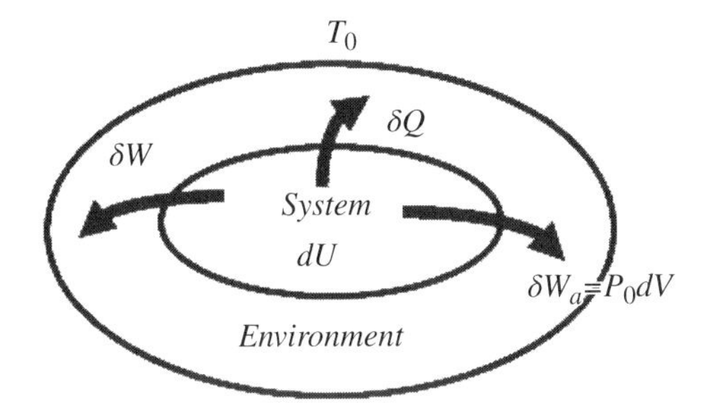

Figure 2.16 The system expands against the atmosphere

Integrating from the initial to final state we have for the actual work

$$W_{12}=\int_1^2 \delta W=\left(U+P_0V-T_0S_{\text{damage}}\right)_1-\left(U+P_0V-T_0S_{\text{damage}}\right)_2-T_0S_{\text{gen}}. \tag{2.80}$$

The lost work or the irreversible work associated with the process is

$$W_{\text{irr}}=T_0S_{\text{gen}}. \tag{2.81}$$

From the second law $S_{\text{gen}}>0$ so that the useful work

$$W_{12}\le\left(U+P_0V-T_0S_{\text{damage}}\right)_1-\left(U+P_0V-T_0S_{\text{damage}}\right)_2=(A_1-A_2). \tag{2.82}$$

S_{gen} is positive from Equation (2.80) or zero, so that

$$W_{12}=\left(\Delta U+P_0\Delta V-T_0\Delta S_{\text{damage}}\right) \tag{2.83}$$

is the available work (free energy). Thermodynamics sometimes refers to terms such as *exergy* or *anergy* (for an isothermal process) or *maximum available work* when the final state of the system is reduced to full equilibrium with its environment [4]. In reliability terms this can be a catastrophic failed state.

Consider the situation where W_{12} has as state 2 the lowest possible energy state so that it is in equilibrium with the environment. This is the maximum availability that gives the maximum useful work. We then take the derivative of this availability A, so with $A=(U+P_0V-T_0S)$

$$\left(\frac{\partial A}{\partial U}\right)_V=0,\quad \left(\frac{\partial A}{\partial V}\right)_U=0, \tag{2.84}$$

which gives

$$1-T_0\left(\frac{\partial S}{\partial U}\right)_V=0,\quad P_0-T_0\left(\frac{\partial S}{\partial V}\right)_U=0. \tag{2.85}$$

Solving, we get the thermodynamic definitions of temperature and pressure:

$$\left(\frac{\partial S}{\partial U}\right)_V=\frac{1}{T_0},\quad \left(\frac{\partial S}{\partial V}\right)_U=\frac{P_0}{T_0}. \tag{2.86}$$

We see that these conditions are satisfied when the system has the same temperature and pressure as the environment at the final state. Thus the largest maximum work is possible when the system ends up in equilibrium with the environment. It is also the maximum possible entropy damage S_D when the final state of the system has catastrophically failed. Then it is impossible to extract further energy as useful work.

In general, one system can provide more work than another similar system. We can assess the irreversible and actual work as

$$W_{irr} = W_{rev} - W_{actual}, \quad W_{actual} = W_{rev} - W_{irr}. \tag{2.87}$$

The efficiency (η) and inefficiency ($1-\eta$) of work-producing components can be assessed as

$$\eta = \frac{W_{actual}}{W_{rev}} = \frac{W_{actual}}{W_{actual} + W_{irr}} \quad \text{and} \quad 1-\eta = \frac{W_{irr}}{W_{actual} + W_{irr}}. \tag{2.88}$$

2.11 Thermodynamic Potentials and Energy States

Thermodynamics is an energy approach. We have held off on introducing the thermodynamic potential until this point as we needed to present some examples related to entropy first. We mentioned briefly in Chapter 1 that free energy is roughly opposite to the entropy. As entropy damage increases for the system, the system's free energy decreases and the system loses its ability to do useful work (see Section 2.5.1). Then instead of focusing on entropy, we can look at the free energy of a system to understand system-level degradation. The thermodynamic potentials are somewhat similar to potential and kinetic energy. When an object is raised to a certain height, it has gained potential energy. When it falls, the potential energy is converted to kinetic energy by the motion. When the object hits the ground, all the potential energy has been converted to kinetic energy. In a similar manner, when we manufacture a device it has a certain potential to do work. This is given by the free energy. When we convert its free energy into useful work, the device's free energy has been converted into useful work that was done. One way to look at it is that free energy comes down to understanding a system's available work, which we will show is one of the most important metrics for products in industry. You might ask: do we really need to understand the thermodynamic potentials to do practical physics? We believe it is beneficial to logically understand thermodynamic degradation and these concepts are used in this book.

There are a number of thermodynamic energy states related to the free energy of a system. The free energy is the internal energy minus any energy that is unavailable to perform useful work. The name "free energy" suggests the energy is available or free with the capacity to do work [18]. So these potentials are useful as they are associated with the internal energy capacity to do work. In this section we will use a combined form of the first and second law for the internal energy, written as

$$dU = TdS + \sum_a Y_a dX_a = TdS - PdV + \sum_i \mu_i dN_i, \tag{2.89}$$

to develop a number of other useful relations called the thermodynamic potentials. Note that since we have substituted $dQ = TdS$, we have assumed a reversible process (see Equation (2.59)) for the moment. The key relations of interest are the Gibbs free energy, Helmholtz free energy, and enthalpy. In the above expression we see that the common natural variables for U are $U = U(S, V)$.

One of the common interesting properties of a system's energy state is assessing its lowest equilibrium state. This is true of the internal energy as well. To look at this consider a system that is receiving heat and doing work $dU = \delta Q - PdV$ (with $dN = 0$, i.e., no particle exchange), then from the second law

$$TdS \geq \delta Q(=PdV + dU) \quad \text{or} \quad TdS - dU \geq dW = PdV. \tag{2.90}$$

Now consider two conditions, constant entropy and constant volume:

$$-(dU)_S \geq dW \quad \text{and} \quad 0 \geq (dU)_{V,S}. \tag{2.91}$$

The subscripts mean that V and S are constant, so dV and dS are zero and Equation (2.61) follows. So for constant system entropy, the work is bounded by the internal energy, that is, the work cannot exceed the free energy. The minus sign can be confusing but if one realizes $U_{\text{initial}} > U_{\text{final}}$ upon integration, the sign goes away. As well when the entropy (S) and the volume of a closed system are held constant, the internal energy (U) decreases to reach a minimum value which is its lowest equilibrium state. This is a spontaneous process. Another way to think of this is that when the internal energy is at a minimum, any increase would not be spontaneous.

2.11.1 The Helmholtz Free Energy

The Helmholtz free energy is the capacity to do mechanical work (useful work). The maximum work is for a reversible process. This is similar to potential energy having the capacity to do work. To develop this potential, we use the chain rule for $d(TS)$ and for PdV work we have:

$$dU = TdS - PdV = d(TS) - SdT - PdV. \tag{2.92}$$

We insert this expression into the internal energy and arrange the equation as

$$d(U - TS) = -SdT - PdV = dF \tag{2.93}$$

where we identify the function $F = U - TS$, known as the Helmholtz free energy. This type of transform is called a Legendre transform; it is a method to change the natural variables. The full expression with particle exchange is

$$dF = -SdT - PdV + \sum_i \mu_i dN_i. \tag{2.94}$$

We see the Helmholtz free energy is a function of the independent variables T, V, denoted $F = F(T, V, N)$.

The change in F (ΔF) is also useful in determining the direction of spontaneous change and evaluating the maximum work that can be obtained in a thermodynamic process. To see this we note that for the first law the total work done by a system with heat added to it also includes a decrease in the internal energy $dU = \delta Q - \delta W$ and since $\delta Q < TdS$ (with the inequality for reversible process) then

$$dU \le TdS - \delta W \tag{2.95}$$

or

$$-\delta W \le dU - TdS = d(U - TS) \text{ when } dT = 0. \tag{2.96}$$

Therefore, when $dT = 0$ we have

$$\delta W \le -dF. \tag{2.97}$$

For an isothermal process ($dT = 0$), the Helmholtz free energy bounds the maximum useful work that can be performed by a system. As the system degrades, the change in the work capability or lost potential work is due to irreversibilities building up in the system; this change in irreversible work resulting from system degradation causes entropy damage that builds up (or is "stored") in the system. We note that W can be any kind of work; it can include electrical ore mechanical (pressure–volume) work including stress strain, that is, not just gases.

The equal sign occurs for a quasistatic reversible process, in which case:

$$\delta W_{\text{reversible}} = \delta W_{\max} = -dF. \tag{2.98}$$

Spontaneous degradation: if we had $\delta W = PdV$ type of work and then decided to hold the volume constant, so that $\delta W = PdV$, Equation (2.68) becomes

$$0 \le -dF, \text{ for } dT = dV = 0 \tag{2.99}$$

where

$$-\int_{\text{initial}}^{\text{final}} dF = F_{\text{initial}} - F_{\text{final}}.$$

We note that when the system spontaneously degrades (losing its ability to perform useful work), the process requires $F_{\text{initial}} \ge F_{\text{final}}$. Note this only requires that the change in the Helmholtz free energy has the same temperature and volume in the initial and final states; along the way the temperature and/or volume may go up or down. This does not necessarily account for the total entropy damage S_{damage} of the system.

We can replace the inequality in Equation (2.97) by adding an irreversible term. We find that when we talk about irreversible processes in Section 2.11.5, for $dT = 0$ the actual work change is

$$-\delta W = -dF + TdS_{\text{damage}}. \tag{2.100}$$

This is an important result as it connects the concept of work, free energy, and entropy damage.

2.11.2 The Enthalpy Energy State

Enthalpy is the capacity to do non-mechanical work plus the capacity to release heat. It is often used in chemical processes. To develop enthalpy, we write PdV using the chain rule

$$PdV = PdV + VdP - VdP = d(PV) - VdP. \tag{2.101}$$

Then we substitute this into the internal energy and manipulate it as

$$dU + d(PV) = TdS + VdP = dH. \tag{2.102}$$

H is called the enthalpy and we can write it as $dH = d(U + PV)$. The full expression for the enthalpy with particle exchange is

$$dH = TdS + VdP + \sum_i \mu_i dN_i. \tag{2.103}$$

We see that enthalpy is a function of the natural variables (S, P), that is, $H = H(S,P,N)$. At constant pressure $dP = 0$ (isobaric) and without particle exchange $dN = 0$; then $\delta Q \leq TdS = H$. This says the enthalpy tells one how much heat is absorbed or released by a system. If for example we have an exothermic reaction where heat flows out of the system to the surroundings, the entropy of the surroundings increases so $\Delta S = -\Delta H / T$. It is common knowledge that a chemical reaction will be spontaneous when the Gibbs free energy (described in the following section) is negative. We mention this because the Gibbs free energy G can be given in terms of the enthalpy $\Delta G = \Delta H - TdS$ which helps explain ΔH. So from this relation, we note that if ΔH is negative and the entropy is increasing than $\Delta G < 0$ and the reaction will proceed.

2.11.3 The Gibbs Free Energy

The Gibbs free energy is the capacity to do non-mechanical work. The maximum work is for a reversible process. This is similar to potential energy having the capacity to do work. To develop the Gibbs free energy we again use the chain rule for $d(PV)$. This time we start by inserting it into the Helmholtz free energy for convenience, obtaining

$$dF = -SdT - PdV = -SdT - d(PV) + VdP \tag{2.104}$$

which can be arranged as

$$d(F + PV) = -SdT + VdP = dG. \tag{2.105}$$

Here we identify the function $G = F + PV = (U - TS) + PV$, which is called the Gibbs free energy. We see the Gibbs free energy is a function of the independent variables $T, P, G = G(T,P)$. Some other relations of interest include $H = U + PV$ or writing it in terms of the enthalpy $G = H - TS$.

We have that $-\delta W \leq dU - TdS$, writing $\delta\text{Work} = \delta W_{\text{other}} + PdV$, where other work refers to non P–V mechanical work such as electrical work. Then at constant temperature and pressure

$$-(PdV+\delta W_{\text{other}})\le dU-TdS \text{ or } -\delta W_{\text{other}}\le dU-TdS+PdV. \tag{2.106}$$

But by definition, $dH = dU + PdV$ so that

$$-\delta W_{\text{other}}\le dH-TdS=dG. \tag{2.107}$$

For an isothermal process at constant pressure, the Gibbs free energy bounds the maximum amount of useful work that can be performed by a system, that is:

$$\delta W_{\text{other}}\le -dG. \tag{2.108}$$

As the system degrades, the change in the work capability or lost potential work is due to irreversibilities building up in the system; this change in irreversible work resulting from system degradation causes entropy damage that is built-up (or "stored") in the system.

In a similar manner to Equation (2.100), we can replace the inequality above by adding the irreversible term

$$\delta W=-dG+TdS_{\text{damage}}. \tag{2.109}$$

If a system is in contact with a heat reservoir meaning constant temperature, and if the pressure is constant, then the change in the Gibbs free energy is

$$\Delta G=\Delta U-T\Delta S-PdV, \text{ or } \Delta S=\frac{\Delta U+P\Delta V-\Delta G}{T}=\frac{Q-\Delta G}{T}. \tag{2.110}$$

The change in the entropy of the system and the reservoir is

$$\Delta S=\Delta S-\Delta S_R=\frac{Q-\Delta G}{T}+\frac{Q_R}{T}=\frac{-\Delta G}{T}, \quad \Delta G\le 0 \text{ (constant} T, P) \tag{2.111}$$

where $Q = -Q_R$. Since ΔS must increase in any process then the Gibbs free energy change must decrease having a negative change in value, $\Delta G \le 0$, when the temperature and pressure are constant in a degradation process.

The Gibbs free energy is a popular function to use when there is particle exchange with the surroundings. In this case, the internal energy is also a function of N so $U = U(S,V,N)$.

From Table 1.1, inserting chemical work, we can write

$$dU=TdS+\delta W=TdS-PdV+\sum_i \mu_i N_i. \tag{2.112}$$

From the above Legendre transform, following the same methods (Equation (2.92)),

$$dG=-SdT-VdP+\sum_i \mu_i N_i. \tag{2.113}$$

Most commonly one considers reactions at constant P and T, so the Gibbs free energy is the most useful potential in studies of chemical reactions as it simplifies to

$$G = \sum_i \mu_i N_i. \tag{2.114}$$

2.11.4 *Summary of Common Thermodynamic State Energies*

From these definitions we can say that ΔU is the energy added to the system, ΔF is the total work done on it, ΔG is the non-mechanical work done on it, and ΔH is the sum of non-mechanical work done on the system and the heat given to it. When we are concerned with the conjugate variables (S, T) and (P, V), we note there are only four possible pairs of independent variables. The only time that we would need to change variables is when we use other types of thermodynamic work besides (P, V).

Table 2.2 summarizes the four types of energies.

2.11.5 *Example 2.13: Work, Entropy Damage, and Free Energy Change*

Let's consider an irreversible process to illustrate some of the ideas in this section with respect to system entropy damage, free energy, and internal energy.

We wish to see how much work can be extracted from a system even during an irreversible process. One way to think of this is to take in heat and see how much work the system can provide. This is illustrated in Figure 2.17. By conservation of energy from the first law for Figure 2.17 we have

$$dU = \delta Q - \delta W. \tag{2.115}$$

Associated with irreversible process will be damage that is occurring in the system (or device). Some of the heat goes into damage (say due to friction) and some into work. Keeping tabs on the system entropy we have

Table 2.2 Four common thermodynamic potential and energy states

Name	Formula	Properties	Derivatives
Internal energy U (S, V) $U = TS - PV$	$dU = \delta Q - \delta W$ $= TdS - PdV$	First law: $\Delta U = Q - W$	$T = \left(\frac{\partial U}{\partial S}\right)_V$, $P = -\left(\frac{\partial U}{\partial V}\right)_S$
Helmholtz free energy $F(T, V)$ $F = U - TS$	$dF = -SdT - PdV$	$\Delta F \leq 0$; T, V constant	$S = -\left(\frac{\partial F}{\partial T}\right)_V$, $P = -\left(\frac{\partial F}{\partial V}\right)_T$
Enthalpy $H(S, P)$ $H = U + PV$	$dH = TdS + VdP$	Isobaric process: $\Delta H = Q$	$T = \left(\frac{\partial H}{\partial S}\right)_P$, $V = \left(\frac{\partial H}{\partial P}\right)_S$
Gibbs free energy $G(T, P)$ $G = U - TS + PV$	$dG = VdP - SdT$ Also of interest: $dG = dH - TdS$	$\Delta G \leq 0$; T, P constant when $G = G(T, P, N)$ and $dG = \sum_i \mu_i dN_i$	$S = -\left(\frac{\partial G}{\partial T}\right)_P$, $V = \left(\frac{\partial G}{\partial P}\right)_Y$

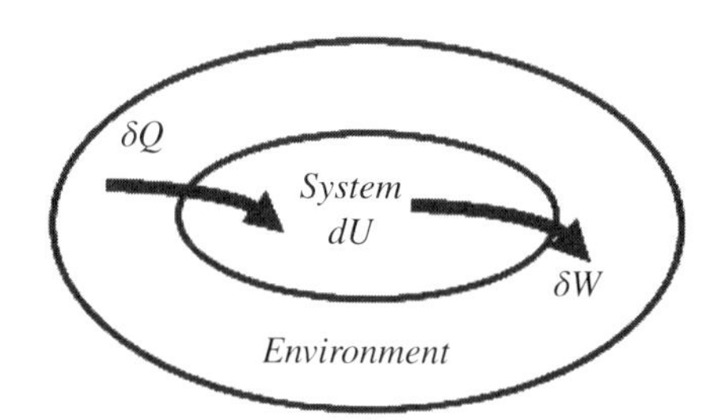

Figure 2.17 Mechanical work done on a system

$$dS_{\text{system}} = dS_{\text{non-damage}} + dS_{\text{damage}} = \frac{\delta Q}{T} + dS_{\text{damage}}. \tag{2.116}$$

Solving for δQ we have

$$\delta Q = TdS_{\text{system}} - TdS_{\text{damage}}, \tag{2.117}$$

and by rearranging terms in Equation (2.115) and inserting the above we obtain

$$-\delta W = dU - \left(TdS_{\text{system}} - TdS_{\text{damage}}\right). \tag{2.118}$$

For the moment let's consider a reversible process so that $dS_{\text{damage}} = 0$. This gives us the maximum work for a reversible process

$$\delta W_{\max} = -dU - TdS_{\text{system}}. \tag{2.119}$$

If $dT = 0$ for an isothermal process we can write this as

$$\delta W_{\max} = \delta W_{\text{rev}} = -d\left(U - TS_{\text{system}}\right) = -dF. \tag{2.120}$$

Therefore, the change in the Helmholtz free energy is defined as the maximum amount of useful work that can be obtained from a system. The work found from the integral going from a system state 1 to 2 in a reversible way is

$$W_{\text{rev},1-2} = -\int_1^2 dF = F_1 - F_2 = \Delta F \quad \text{where} \quad F_1 > F_2. \tag{2.121}$$

Now what about the irreversible process? We need to somehow account for what happens to the system's free energy. When in state 1 we know that its free energy is F_1. In that state it had a certain capacity to do work which was bounded by the free energy F_1, so the work capability was $W \le F_1$.

The path-dependent work causes resistance or friction and the actual work in going to state 2 was really

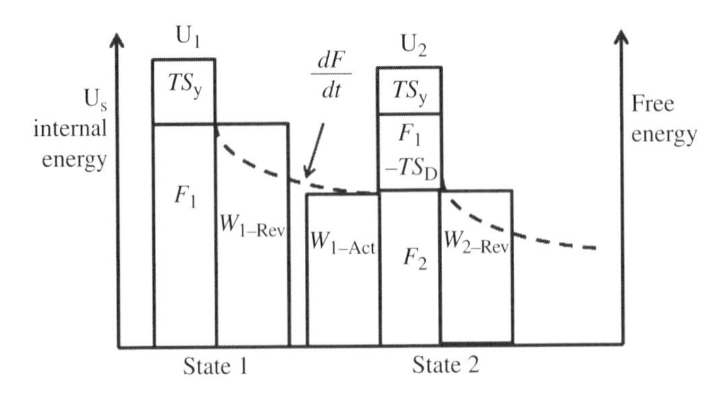

Figure 2.18 Loss of available work due to increase in entropy damage

$$W_{\text{actual},\,1-2} = F_1 - T\Delta S_{\text{damage},\,1-2} = W_{\text{rev},\,1} - T\Delta S_{\text{damage},\,1-2} = F_2. \tag{2.122}$$

So now state 2 has a different free energy then state 1 due to the entropy damage. Note that we have dropped the minus sign for the work as it is understood we seek the useful work that a system does.

Now if we continued to do work to a final equilibrium state F_N in an incremental way we would get to the minimum value of the free energy F_{min}, but over the same path we would have

$$F_1 > F_2 > F_3 > \ldots > F_N = F_{\text{min}}. \tag{2.123}$$

We can visualize this incremental process with the aid of the Figure 2.18 for a system that has undergone an irreversible process and incurred entropy damage between state 1 and state 2. There is no way to reverse the entropy damage. According to the second law, $dS > 0$ for the irreversible process and entropy increase means the system disorder has increased. If state 2 happens to be its lowest energy state then no further work can occur; the system has likely failed catastrophically so entropy damage stops, $dS_{\text{damage},2} = 0$, and the change in the free energy goes to zero $\Delta F = 0$.

When we talk about $dT = 0$ it is more appropriate to say that only the initial and final states need to have the same temperature. In between they can fluctuate. We can write the incremental change in the work as

$$\delta W_{\text{actual}} = F_{\text{initial}} - TdS_{\text{damage}} \quad \text{or} \quad \delta W_{\text{actual}} = W_{\text{rev}} - TdS_{\text{damage}} \tag{2.124}$$

or

$$W_{\text{actual}} = W_{\text{rev}} - W_{\text{irr}}. \tag{2.125}$$

This is consistent with Equation (2.87). The efficiency of the work process is then

$$\eta = \frac{W_{\text{actual}}}{W_{\text{rev}}}. \tag{2.126}$$

At this point we can start modeling the actual work and the irreversible work. For example we might write

$$W(t)_{\text{actual}} = W_{\text{rev}} - W(t)_{\text{irr}}. \tag{2.127}$$

Since the reversible work (the maximum work) can be treated as fixed, the rate of actual work is some function of the rate of the irreversible work over the work path

$$\frac{\delta W_{\text{actual}}}{dt} = \frac{\delta W_{\text{irr}}}{dt}. \tag{2.128}$$

These last equations seem harmless enough but they are important results. These work functions are independent of whether we talk about a chemical process where we are using the Gibbs free energy, an electric, or a mechanical process using the Helmholtz free energy. If you are a pragmatic person, the last few equations give you a great bottom line for thermodynamics because no matter how complicated thermodynamics is, a big part of it is undeniably all about *work.*

For this problem the general expression for power is:

$$-\frac{\delta W}{dt} = -\frac{dF}{dt} + T\frac{dS_{\text{damage}}}{dt} = \text{Power loss.} \tag{2.129}$$

Now the work left that is available in state 2 can be more time related than power loss related. That is, the system has lost some strength from the irreversible process U_{12}; however, it may very well be able to keep on performing its intended function until its free energy is all used up. We provided an example in Section 2.1 on Miner's rule (see also Chapter 4) that is commonly used to interpret the work left in terms of cycles of remaining (time) and not power or strength loss.

We also note that the entropy damage is path dependent on how the work was performed. As we mentioned in Chapter 1 we are staying away from using too many confusing symbols but, in reality, we should be mindful about the path dependence related to the work path for which we should be writing dS_{damage} and $dS_{\text{non-damage}}$ as δS_{damage} and $\delta S_{\text{non-damage}}$.

2.11.6 *Example 2.14: System in Contact with a Reservoir*

Consider a special case of a system that is in contact with a heat reservoir, as shown in Figure 2.19. If there is no mechanical work $PdV = 0$ with constant volume $\Delta U = \Delta Q$. Then

$$\Delta F = \Delta U - T\Delta S_{\text{system}} \text{ so that } \Delta S_{\text{system}} = \frac{\Delta U_S - \Delta F}{T} = \frac{\Delta Q - \Delta F}{T}. \tag{2.130}$$

Then the total entropy change is

$$\Delta S = \Delta S_{\text{system}} + \Delta S_R = \frac{\Delta Q - \Delta F}{T} + \frac{\Delta Q_R}{T} = -\frac{\Delta F}{T} \tag{2.131}$$

where $\Delta Q_S = -\Delta Q_{\text{reservoir}}$.

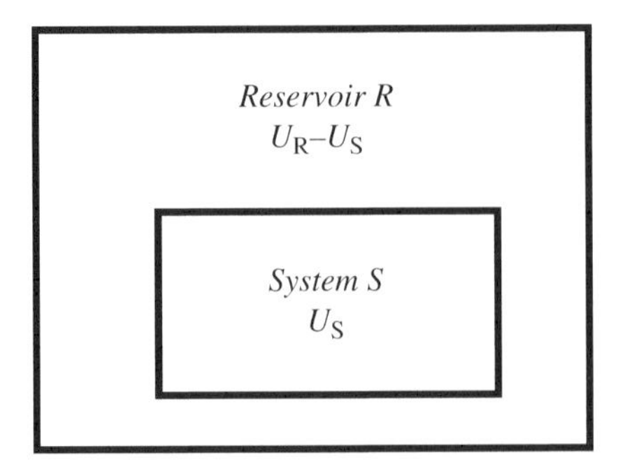

Figure 2.19 A simple system in contact with a heat reservoir

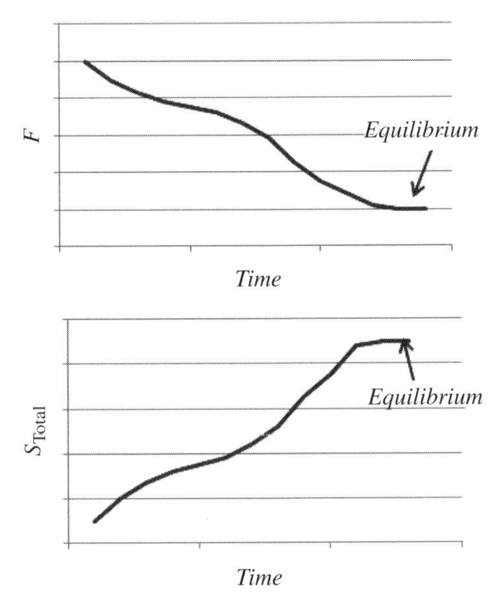

Figure 2.20 A system's free energy decrease over time and the corresponding total entropy increase

Note that the total entropy $\Delta S > 0$ so we can also write

$$\Delta F \leq 0 \quad \text{for} \quad dT = dV = 0. \tag{2.132}$$

The Helmholtz free energy never increases in a spontaneous process with fixed temperature and volume in a system in contact with a heat reservoir. We see this causes the entropy to increase since $\Delta S = -\Delta F/T$. To emphasize this, as the system's entropy increases, its free energy will have a tendency to decrease if it can. Figure 2.20 shows the possible spontaneous decay of a system's free energy over time to its final descent to its equilibrium state and the corresponding total entropy increase.

We note that we have sketched the free energy which decays in a non-linear way. This is not atypical of systems.

2.11.6.1 Full Expression for Potential with Entropy Damage

General work-related processes and exchanges with the system's internal energy require different relationships. For example, consider the full assessment without any conditions on the Helmholtz free energy and temperature where particles can flow in or out of the system; then the general change in the work is

$$\delta W = -dF + TdS_{\text{damage}} - SdT + \sum_i \mu_i dN_i \tag{2.133}$$

where N_i is the number of particles or moles in the system of species i. Note in Section 2.11.1 we used conditions $dT = dN = 0$. One needs to account for sign changes depending on the particular conditions as well.

If we have a situation of constant temperature and pressure ($dT = dP = 0$) we will obtain the similar results in terms of the Gibbs free energy

$$W_{12} = -\Delta G + T\Delta S_{\text{damage}}. \tag{2.134}$$

The general assessment without any conditions on the Gibbs free energy is

$$\delta W = -dG + TdS_{\text{damage}} - SdT + VdP + \sum_i \mu_i dN_i. \tag{2.135}$$

If we have a situation of constant pressure and entropy flow ($dT = dS = 0$) we will obtain similar results in terms of the enthalpy:

$$W_{12} = -\Delta H + T\Delta S_{\text{damage}}. \tag{2.136}$$

The general assessment without any conditions on the enthalpy is

$$\delta W = -dH + TdS_{\text{damage}} - TdS + VdP + \sum_i \mu_i dN_i. \tag{2.137}$$

Lastly, the internal energy is

$$\delta W = -dU + TdS_{\text{damage}} - TdS + \sum_i \mu_i dN_i. \tag{2.138}$$

Summary

2.1 Cumulative Entropy Damage Approach in Physics of Failure

The total entropy of a system is equal to the sum of the entropies of the subsystems (or parts), so the entropy change is

$$\Delta S_{\text{system}} = \sum_{i}^{N} \left(\Delta S_{\text{subsystem}}\right)_i. \tag{2.1}$$

We now can define here the Cumulative Entropy Damage Equation:

$$\Delta S_{\text{system-cum-damage}} = \sum_{i}^{N} \left(\Delta S_{\text{subsystem-damage}}\right)_i. \tag{2.2}$$

This seemingly obvious outcome (Equation (2.2)) is an important result not only for system reliability but also for device reliability.

The rate-controlling failure part in a complex system can be detected if its resolution is reasonable where

$$i\text{th entropy damage resolution} = \frac{\Delta S_{i,\text{damage}}}{\Delta S_{\text{system}}}.$$

2.1.1 *Example 2.1: Miner's Rule Derivation*

Consider that we have M total cells of potential damage to failure that would lead to a maximum cumulative damage $S_{\max}$. Then the cumulative damage entropy is simply

$$\text{Cum Damage} = \frac{\sum_n s_{n,\text{damaged}}}{S_{\max}} = \frac{\sum_m m_{\text{damaged-cells}}}{M} \le 1.$$

Note that if N is the number of cycles to failure and n is the number of cycles performed ($N > n$) then we can write for the ith stress level

$$\text{Cum Damage} = \frac{n_1}{N_1} + \frac{n_2}{N_2} + \ldots + \frac{n_k}{N_k} = \sum_{i=1}^{K} \frac{n_i}{N_i}. \tag{2.8}$$

This equation is called Miner's rule. It follows that we can also invent here a related term called

$$\text{Cumulative Strength} = 1 - \text{Damage}. \tag{2.9}$$

2.1.3 Non-Cyclic Applications of Cumulative Damage

We noted that we could write, similar to Miner's Rule, cumulative damage for non-cyclic time-dependent processes as

$$\text{Cum Damage} = \sum_{i=1}^{K} \left(\frac{t_i}{\tau_i}\right)^P. \tag{2.10}$$

As in the Miner's cyclic rule, by analogy t_i is the time of exposure of the ith stress level and τ_i is the total time to cause failure at the ith stress level. Since not all processes age linearly with time, as is found in Chapter 4, the exponent P was added for the generalized case where P is different from unity.

2.2 Measuring Entropy Damage Processes

To measure entropy damage, we consider a measurement process f that must be consistent to a point that it is repeatable at a much later aging time t_2. If some measurable degradation has occurred, the entropy change is

$$\Delta S_f(t_2) = S(t_2 + \Delta t) - S(t_2), \text{ where } t_2 >> t. \tag{2.13}$$

If our measurement process f at time t_1 and t_2 is consistent, we should find the entropy damage that has occurred between these measurement times is

$$\Delta S_{f,\text{damage}}(t_2, t_1) = \Delta S_f(t_2) - \Delta S_f(t_1) \geq 0. \tag{2.14}$$

2.4.2 Example 2.3: Resistor Aging

Resistor aging is a fundamental example, since resistance generates entropy. Entropy damage change for resister degradation was found to be

$$\Delta S_f(t_1) = mC_{p,\text{avg}} \ln \frac{T_3}{T_2} \tag{2.19}$$

or, in terms of an aging ratio,

$$A_{\text{aging-ratio}} = \ln(T_3/T_2)/\ln(T_4/T_2). \tag{2.23}$$

Simple resistance change was then written as

$$R(t_f) = \left(\frac{T_4 - T_2}{T_3 - T_2}\right) R(t_i). \tag{2.24}$$

2.4.3 Example 2.4: Complex Resistor Bank

For a complex bank, if we are able to use the exact same measurement process f as for Section 2.4.2 then the aging ratio is still

$$A_{\text{aging-ratio}} = \ln(T_3/T_2)/\ln(T_4/T_2). \tag{2.27}$$

2.4.4 System Entropy Damage with Temperature Observations

For a system made up of similar-type parts, in accordance with Equation (2.28) the total entropy damage is

$$\Delta S_{\text{total}} = \sum_{i=1}^{N} \Delta S_i = \ln\frac{T_2}{T_1}\sum_{i=1}^{N} m_i C_{\text{avg},\,i}. \tag{2.29}$$

2.5 Measuring Randomness due to System Entropy Damage with Mesoscopic Noise Analysis in an Operating System

We defined an operating system here as follows.

An operating system has a thermodynamic process that generates energy (battery, engine, electric motor...) or has energy passing through the system (transistor, resistor, capacitor, radio, amplifier, human heart, cyclic fatigue...).

We define an Entropy Noise Principle as follows.

In an aging system, we suspect that disorder in the system can introduce some sort of randomness in an operating system process leading to system noise increase. This we suspect is a sign of disorder and increasing entropy. Simply put, if entropy damage increases, so should the system noise in the operating process.

Then we defined a mesoscopic noise entropy measurement as follows.

A mesoscopic noise entropy measurement is any measurement method that can be used to observe subtle system-level entropy damage changes that generate noise in the operating system.

To understand the concept of noise the statistical definition of entropy was introduced as we needed to deal with continuous random variables:

$$\text{Continuous}\,X, f(x):\ \ S(X) = -\int f(x)\log_2(f(x))dx = -E[\log f(x)]. \tag{2.33}$$

2.5.1 *Example 2.6: Gaussian Noise Vibration Damage*

The differential entropy equation for Gaussian noise system is

$$S(X)=\frac{1}{2}\log\left[2\pi e\sigma(x)^2\right]. \tag{2.35}$$

We see that for Gaussian type noise, the deferential entropy is only a function of its variance σ^2 (it is independent of its mean μ). We noted that this is an important result for system-level degradation analysis. For a system that is degrading by becoming noisier over time, the entropy damage can be measured in a number of ways where the change in the entropy at two different times t_2 and t_1 is

$$\Delta S_{\text{damage}}=S_{t2}(X)-S_{t1}(X)=\Delta S(t_2,t_1)=\frac{1}{2}\log\left(\frac{\sigma_{t2}^2}{\sigma_{t1}^2}\right). \tag{2.36}$$

2.6 How System Entropy Damage Leads to Random Processes

We defined a System Entropy Random Process Postulate:

Entropy increase in a system due to aging causes an increase in its disorder which introduces randomness to thermodynamic processes between the system and the environment.

Then we noted a System Entropy Damage and Randomness in Thermodynamic State Variables Postulate.

Entropy damage in the system likely introduces some measure of randomness in its state variables that are measurable when energy is exchanged between the system and the environment.

This problem reduces to finding what we refer to as the normalized Noise Entropy Autocorrelation Function:

$$R_{s,s}(\tau)_{\text{norm}}=\frac{E[S(t),S(t+\tau)]}{E[S(t),S(t)]}=\frac{\int_{-\infty}^{\infty}S(t)S(t+\tau)dt}{\int_{-\infty}^{\infty}S(t)^2dt}, \tag{2.51}$$

where

$$-1\leq R_{s,s}(\tau)_{Norm}\leq 1$$

and $R_{S,S}(\tau)$ is the entropy noise autocorrelation function and E indicates that we are taking the expectation value.

Here $E[S(t), S(t)]$ is the maximum value. We see that any thermodynamic variable that is part of the entropy exchange process between the environment and the system is subjected to randomness stemming from system aging disorder of which one measurement type is to look at its autocorrelation function.

2.6.1 Stationary versus Non-Stationary Entropy Process

Measured data can often be what is termed non-stationary, which indicates that the process varies in time, say, with its mean and/or variances that change over time. Therefore, non-stationary data are typically unpredictable and hard to model for autocorrelated analysis. Although aging processes are non-stationary, they may vary slowly enough so that in the typical measurement time frame, they are stationary.

2.7 Example 2.8: Human Heart Rate Noise Degradation

Although noise degradation measurements are difficult to find, one helpful example of noise autocorrelation analysis of an aging system found by the author is in an article by Wu *et al.* [10] on the human heart. Here heart rate variability was studied in young, elderly, and CHF patients. The study demonstrated that heart rate noise between young and elderly patients is not dramatically different compared to what is occurring in patients with CHF. This demonstrated the power to detect subtle damage.

2.8 Entropy Damage Noise Assessment Using Autocorrelation and the Power Spectral Density

The autocorrelation function in the time domain

$$R_{yy}(\tau) = \int_{-\infty}^{\infty} y(t)y(t+\tau)\,dt \tag{2.52}$$

could be transformed to the frequency domain. The Fourier transform of the autocorrelation function provides what is termed the PSD Spectrum given by

$$PSD(f) = S_{yy}(f) = \int_{-\infty}^{\infty} R_{yy}(\tau)e^{-j2\pi f\tau}\,d\tau. \tag{2.54}$$

2.8.1 Noise Measurements Rules of Thumb for the PSD and R

1. As the absolute value of the autocorrelation function decreases to zero, $|R_{norm}| \to 0$, noise levels are increasing in the system.
2. As the PSD(f) levels increase, noise is increasing in the system.

3. The slope of the autocorrelation function is related to the slope of the PSD spectrum.
4. As the autocorrelation slope increases, the PSD slope decreases.
5. The slope is an indication of the type of system-level noise characteristic detailed in the next section.

2.8.2 Literature Review of Traditional Noise Measurement

Noise is sometimes characterized with color [11]; the three main types in this category are white, pink, and brown noise. Each has a specific type of broadband PSD characteristic. We will first overview these then we will describe some reliability-related type of noise measurements. Each type has an interesting sound that is audible in the human audio frequency range (~20 Hz to 20 kHz) area of the spectrum. The PSD(f) spectral characteristic of these have different slopes: white noise has a constant slope; pink noise slope varies with $1/f$; and brown noise varies with $1/f^2$.

2.8.3 Literature Review for Resistor Noise

Resistance is a type of friction which increases temperature, creates heat, and often entropy damage. In the absence of friction, the change in entropy is zero. Therefore, it makes sense in our discussion on noise as an entropy indicator that we look at noise in electrical resistors.

Studies showed that resistors with reduced volume or cross-sectional area produce more noise. Carbon resistor noise type was $1/f$ type (pink noise).

2.9 Noise Detection Measurement System

A noise detection measurement system that used the autocorrelation method was described for an operating or non-operating system. We feel that an active on-board autocorrelation measurement can be helpful for prognostics in determining a system's wellness in real time.

2.10 Entropy Maximize Principle: Combined First and Second Law

We may combine the general first law statement

$$dU = \delta Q + \delta W = \delta Q + \sum_a Y_a dX_a \tag{2.59}$$

with the statement of the thermodynamic second law $\delta Q = TdS$ (for reversible processes). For a general system the combined statement reads for a reversible process

$$dU = TdS + \sum_a Y_a dX_a . \tag{2.60}$$

The entropy maximum principle in terms of aging: given a set of physical constraints, a system will be in equilibrium with the environment if and only if the total entropy of the system and environment is at a maximum value.

$$S_{\text{total}} = S_{\max} = S_{\text{system}} + S_{\text{env}}. \tag{2.62}$$

At this point, the system has lost all its useful work, maximum damage has occurred where $dS_{\text{total}} = 0$, and no further aging will occur.

2.10.1 Example 2.9: Thermal Equilibrium

If the entropy is at the maximum value then the first order differential vanishes. For a thermodynamic system shown in Figure 2.14, the thermodynamic equation related to the equilibrium condition was

$$dS_{\text{total}} = \left(\frac{1}{T} - \frac{1}{T_{\text{env}}}\right) dU + \left(\frac{P}{T} - \frac{P_{\text{env}}}{T_{\text{env}}}\right) dV. \tag{2.65}$$

At equilibrium the energy exchange will come to a halt when the environmental temperature is equal to the system temperature $T = T_{\text{env}}$, and $P = P_{\text{env}}$.

2.10.2 Example 2.10: Equilibrium with Charge Exchange

The combined first and second law equation for the battery–capacitor in Figure 1.6 is

$$dS_{\text{total}} = \left(\frac{1}{T} - \frac{1}{T_{\text{env}}}\right) dU - \left(\frac{V}{T} - \frac{E}{T_{\text{env}}}\right) dq. \tag{2.75}$$

For equilibrium when the entropy is maximum, $T = T_{\text{env}}$ and $V = E$, so that $dS_{\text{total}} = 0$, the battery has in a sense degraded its energy in charging the capacitor so that, in the absence of capacitor leakage, no more useful work will occur.

2.11 Thermodynamic Potentials and Energy States

Thermodynamics is an energy approach. As entropy damage increases for the system, the system's free energy decreases and the system loses its ability to do useful work. Then instead of focusing on entropy, we can look at the free energy of a system to understand system-level degradation. This comes down to understanding system work, which we will show is one of the most important metrics for products in industry.

The free energy is the internal energy minus any energy that is unavailable to perform useful work. The name "free energy" suggests the energy is available or free with the capacity to do work. So these potentials are useful as they are associated with the internal energy capacity to do work. The key relations of interest are the Gibbs free energy, Helmholtz free energy, and enthalpy.

The work in terms of the Helmholtz free energy F is

$$\delta W = -dF + TdS_{\text{damage}} - SdT + \sum_i \mu_i dN_i. \quad (2.133)$$

The general assessment without any conditions on the work in terms of the Gibbs free energy G is

$$\delta W = -dG + TdS_{\text{damage}} - SdT + VdP + \sum_i \mu_i dN_i. \quad (2.135)$$

The general assessment without any conditions for the work in terms of the enthalpy H is

$$\delta W = -dH + TdS_{\text{damage}} - TdS + VdP + \sum_i \mu_i dN_i. \quad (2.137)$$

Lastly, in terms of the internal energy U without any conditions for the work:

$$\delta W = -dU + TdS_{\text{damage}} - TdS + \sum_i \mu_i dN_i. \quad (2.138)$$

We noted from the free energy change that we could write the work in terms of actual, reversible, and irreversible processes such that

$$W_{\text{actual}} = r_{\text{rev}} - W_{\text{irr}}. \quad (2.125)$$

This is consistent with Equation (2.88). The efficiency of the work process is then

$$\eta = \frac{W_{\text{actual}}}{W_{\text{rev}}}. \quad (2.126)$$

Since the reversible work, the maximum work can be treated as fixed; the rate of actual work is some function of the rate of the irreversible work path:

$$\frac{\delta W_{\text{actual}}}{dt} = \frac{\delta W_{\text{irr}}}{dt}. \quad (2.128)$$

These work equations are important results. It is apparent that any work process independent of whether we talk about a chemical process, the Gibbs free energy, or a mechanical process using the Helmholtz free energy, the bottom line is undeniably that it is all about *work*.

References

[1] Miner, M.A. (1945) Cumulative damage in fatigue. *Journal of Applied Mechanics*, **12**, A159–A164.
[2] Feinberg, A. (2014) Thermodynamic damage within physics of degradation, in *The Physics of Degradation in Engineered Materials and Devices* (ed J. Swingler), Momentum Press, New York.
[3] Shannon, C.E. (1948) A mathematical theory of communication. *Bell System Technical Journal*, **27** (3), 379–423.
[4] Hazewinkel, M. (ed) (2001) Differential entropy, in *Encyclopedia of Mathematics*, Springer, New York.
[5] Allan, D. (1966) Statistics of atomic frequency standards. *Proceedings of IEEE*, **54** (2), 221–230.
[6] Feinberg A. Thermodynamic damage measurements of an operating system. Proceedings of the 2015 Annual Reliability and Maintainability Symposium (RAMS), January 26–29, 2015, Palm Harbor, FL. IEEE Xplore.
[7] Feinberg A. Active autocorrelation noise detection degradation spectroscopy library measurement device. US Patent 62/284,056, Sept 21, 2015. (For prognostic noise measurement software tools, interested readers should contact the author through his website http://www.dfrsoft.com.)
[8] Buckingham, M.J. (1985) *Noise in Electronic Devices and Systems*, John Wiley & Sons, Inc, New York.
[9] Dunn, P.F. (2005) *Measurement and Data Analysis for Engineering and Science*, McGraw–Hill, New York.
[10] Wu, G.Q., Arzeno, N.M., Shen, L.L., Tang, D.K., Zheng, D.A., Zhao, N.Q., Eckberg, D.L. Chaotic signatures of heart rate variability and its power spectrum in health, aging and heart failure. *PLoS One*,2009, **4** (2), e4323.
[11] Musha, T., Sato, S. and Yamamot, M. (eds) (1992) *Noise in Physical Systems and 1/f Fluctuations*, IOS Press, Amsterdam.
[12] Keshner, M.S. (1982) 1/f noise. *Proceedings of the IEEE*, **70** (3), 212–218.
[13] Caloyannides, M.A. (1974) Microcycle spectral estimates of l/f noise in semiconductors. *Journal of Applied Physics*, **45** (1), 307–316.
[14] Mandelbrot, B.B. and Wallis, J.R. (1969) Some long-run properties of geophysical records. *Water Resources Research*, **5** (2), 321–340.
[15] Barry, P., Errede, S. Measurement of 1/f noise in carbon composition and thick film resistors. Senior thesis, Phys 499, Fall 2014, University of Illinois.
[16] de Jeu, W.H., Geuskens, R.W.J. and Pike, G.E. (1981) Conduction mechanisms and 1/f noise in thick-film resistors with Pb3Rh7015 and PbQRU207. *Journal of Applied Physics*, **52**, 4128.
[17] Wolf, M., Muller, F. and Hemschik, H. (1985) Active Passive. *Electronic Components*, **12**, 59.
[18] Morse, P.M. (1969) *Thermal Physics*, Benjamin/Cummings Publishing, New York.

3

NE Thermodynamic Degradation Science Assessment Using the Work Concept

3.1 Equilibrium versus Non-Equilibrium Aging Approach

We briefly discussed equilibrium versus non-equilibrium (NE) thermodynamics damage in Section 1.3.1. Here we will use an energy approach to physics of failure problems. In NE thermodynamic degradation, we are concerned about how the aging process takes place over time so that it can be modeled. In a sense we already have one approach to this. In Chapter 2, we devised a consistent measurement process f to help measure the damage that may have occurred between times t_1 and t_2, finding the entropy damage was, as given by Equation (2.14):

$$\Delta S_{f,\text{damage}}(t_2, t_1) = \Delta S_f(t_2) - \Delta S_f(t_1) \geq 0 \tag{3.1}$$

where $\Delta S_f(t_i)$ is a quasistatic measurement. Here we can add as many intermediate quasistatic thermodynamics measurements as needed to trace out the aging path. Each measurement is taken when the system is either under very little stress or over a short enough time period, so that we are able to sample the system's state variables over time. We then trace out how the damage is evolving over an extended time period. In this way we will be able to model the aging that is occurring by fitting the data. However, using this method we need to keep track and measure a number of states throughout the aging process. Often we are able to model the degradation (damage) so we do not have to make many measurements. In this chapter, we set the stage for NE thermodynamic damage assessment. Most of the rest of this book deals with NE thermodynamics science methods.

Thermodynamic Degradation Science: Physics of Failure, Accelerated Testing, Fatigue, and Reliability Applications, First Edition. Alec Feinberg.

3.1.1 *Conjugate Work and Free Energy Approach to Understanding Non-Equilibrium Thermodynamic Degradation*

An approach for understanding the type of thermodynamic stresses that are occurring to the aging system is to use the conjugate work approach that we established in Section 1.4. During the quasistatic process, the work done on the system by the environment or by the system on the environment can be assessed using the conjugate variables listed in Table 1.1:

$$\delta W = \sum_a Y_a dX_a \text{ or } W = \int_{X_1}^{X_2} Y\, dX \tag{3.2}$$

where we need to sum the work. If the path is well defined, we can integrate over the path from X_1 to X_2. Each generalized displacement dX_a is accompanied by a generalized conjugate force Y_a. Recall that because work is a function of how it is performed (often termed path dependent in thermodynamics), we used the symbol δW instead of dW indicating this. For a simple system there is but one displacement X accompanied by one conjugate force Y as given in Table 1.1.

We discussed how thermodynamics is an energy approach in Section 1.1 and Chapter 2. As entropy damage increases for the system, the system's free energy decreases and the system loses its ability to do useful work (see Equation (2.131) and Figure 2.20). Then instead of focusing on entropy, we can look at the free energy of a system to understand system-level degradation. This comes down to understanding system work which we will show is one of the most important metrics for products in industry. For example, from Chapter 2 we noted that the work term for the Helmholtz free energy can be written

$$\delta W = -dF + TdS_{\text{damage}} \tag{3.3}$$

where, for simplicity, we used the conditions that $dT = dN = 0$. We also note that for isothermal actual work, it is bounded by the free energy $W \leq \Delta F$. For a reversible process, the equality indicates that the free energy is the reversible work. Equation (3.3) can then be interpreted as (see also Section 2.10.4):

$$W_{\text{actual}} = W_{\text{rev}} - W_{\text{irr}}. \tag{3.4}$$

This is the thermodynamic work. Although this was obtained with simplified specific Helmholtz isothermal conditions, it can be found for the Gibbs free energy (see Chapter 5) and it proves to be relevant to many thermodynamic work processes as it is a general transparent statement. It is a statement about the free energy that equals the reversible work, as it is considered the maximum useful available work of the system.

The thermodynamic work is perhaps the most directly measurable and practical quantity to use in assessing the system's free energy.

Therefore, we see that the actual work really characterizes the situation for many system processes. It is a good indication of the change in the system's ability to do useful work and the damage that is created along the work path.

We can treat the initial value for the reversible work (the maximum work) prior to introducing irreversibilities as a fixed value. Then the change in actual work is some function of the change of the irreversible work over the work path:

$$\delta W_{\text{actual}} = \delta W_{\text{irr}}. \tag{3.5}$$

Many problems in non-equilibrium thermodynamics which can be put in the category of physics of failure for semiconductors, fatigue, creep, wear, etc. come down to assessing the actual work which is dependent on the irreversible work. The irreversible work is of course dependent on the entropy damage occurring along the work path.

Therefore, in this chapter, we will explore non-equilibrium degradation processes by assessing the thermodynamic work.

3.2 Application to Cyclic Work and Cumulative Damage

In cyclic reversible work the system undergoes a process in which the initial and final system states are identical. A simple example for non-reversible cyclic work is the bending of a paper clip back and forth. The cyclic thermodynamic work is converted into heat and entropy damage in the system as dislocations are added every cycle, causing plastic deformation. This produces metal fatigue and irreversible damage, as illustrated in Figure 3.1. Aging from such fatigue is due to external forces, which eventually result in fracture of the paper clip.

If we were trying to estimate the amount of damage done each cycle, it might be more accurate to write the damage in terms of the number of dislocations produced during each cycle (see Section 2.1.1). Then we could also estimate the cumulative damage, although this number is usually unknown. We therefore take a thermodynamic approach.

For a simple system, the path in the (X, Y) plane corresponding to a cycle is shown in Figure 3.2.

The work term is $\delta W_s = YdX$. Recall that the plane can represent any one of the conjugate work variables given in Table 1.1 that can undergo cyclic work. In the (X, Y) plane, the cycle

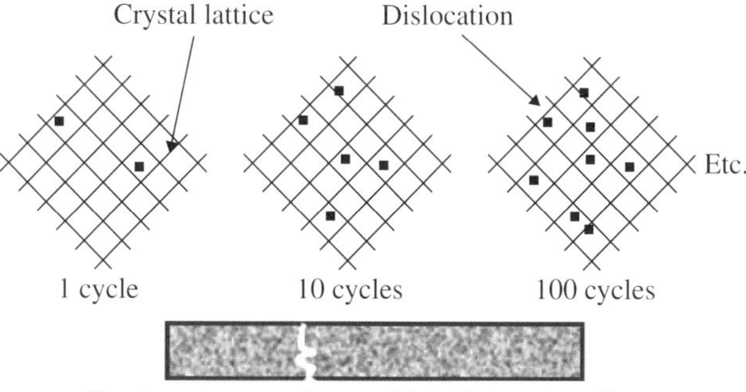

Figure 3.1 Conceptual view of cyclic cumulative damage. Source: Feinberg and Widom [1]. Reproduced with permission of IEEE

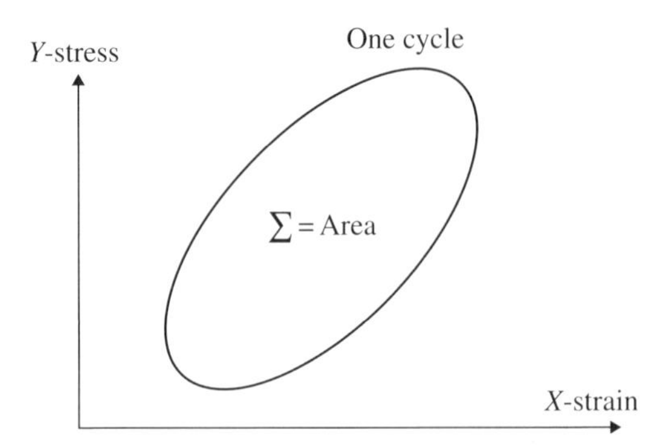

Figure 3.2 Cyclic work plane

is represented by a closed curve C enclosing an area Σ. The closed curve is parameterized during the time interval $t_{\text{initial}} < t < t_{\text{final}}$ of the cycle as a moving point $(X(t), Y(t))$ in the plane.

For a cyclic process C in a simple system, the work done on the system by the environment is given by the area obtained from the integral around a closed curve, that is:

$$\text{Work}(\text{cycle}) = \oint_C Y dX = \int_{t_{\text{initial}}}^{t_{\text{final}}} Y(t) \frac{dX(t)}{dt} dt. \tag{3.6}$$

The curve C representing the cycle is the boundary of an area Σ which is written here as $C = \partial\Sigma$. Employing Stokes integral theorem, one proves that the work done during a cycle may be related to the enclosed area via

$$\text{Work}(\text{cycle}) = \oint_{\partial\Sigma} Y dX = \iint_{\Sigma} dY dX. \tag{3.7}$$

In Figure 3.2, the system does work on the environment if the point in the plane transverses the curve in a clockwise fashion. The environment does work on the system if the point in the plane transverses the curve in a counterclockwise fashion.

A system undergoing cycles acts as an *engine* if the system does work on the environment during each cycle. The system acts as a *refrigerator* if the environment does work on the system during each cycle. In either case, during a cyclic process the system is restored to its initial state. Perhaps unfortunately, the environment is not restored to its environmental initial state after the completion of a cycle. The environment undergoes some *damage* during each cycle in terms of *entropy* in accordance with the second law of thermodynamics.

Consider a system with matter inside of automotive cylinders in a typical automobile engine. If the engine is firing on all cylinders, then the number of cycles per second would be best measured by the tachometer in terms of revolutions per minute (RPM). Before and after each cycle of motion for these cylinders, the chemical content of each cylinder will be the same (replenished).

Let us say that each cycle starts with a given mixture of petroleum vapor and air inside each of the cylinders. During the cycle, the petroleum vapor and air chemically react (fuel burning) and the end-products of this chemical reaction will be belched out of the cylinders. Eventually, the noxious fumes exit the automobile through the exhaust pipe. The cylinders will move up and down exactly once during each cycle with some resulting pressure–volume cylinder work. Perhaps the environmental cylinder work (in part) slowly drags the automobile up a very annoying steep hill. A fresh new mixture of petroleum vapor and air is sucked into the cylinders at the end of each cycle, ending the completed old cycle and beginning a fresh new cycle.

A car owner might periodically ask a mechanic, in the language of American slang, "what is the *damage* on my automobile?" In thermodynamics we might ask, what is the damage caused to the environment during a system cycle as measured in entropy units. Here, we can discuss the notion of damage qualitatively keeping the common cyclic automobile engine example in our minds. There is damage to the environmental atmosphere when the chemical output of the fuel burning is belched out of the exhaust pipe. There is also environmental damage to the cylinders from the constant banging of the cylinder heads up and down during each cycle. Recall that the system chemical contents of a cylinder give rise to a pressure–volume work term $\delta W = -PdV$. The cylinder wall and the cylinder head are part of the environment. The volume V inside the cylinder is an *external* parameter. Environmental "entropy damage" to the engine cylinder heads may get converted into the "financial damage" required for rebuilding the engine. Automobile engine failure is a typical example of the environmental entropy damage present in any real cyclic thermodynamic engine.

3.3 Cyclic Work Process, Heat Engines, and the Carnot Cycle

We typically think of an engine as a device that converts heat into motion, such as an automobile gas engine or a steam engine. We can start by considering a quasistatic cyclic process shown in Figure 3.3. The "oval" curve is an arbitrary cyclic process and the dotted lines represent a Carnot cyclic process. A Carnot cycle is often described in most thermodynamic text books because it is the most efficient heat engine that is allowed by the laws of thermodynamics. The Carnot cycle is a quasistatic operation and thus operates too slowly to be practical. However, we can use it as a good overview to illustrate the thermodynamic cyclic work and also discuss entropy damage to a heat engine.

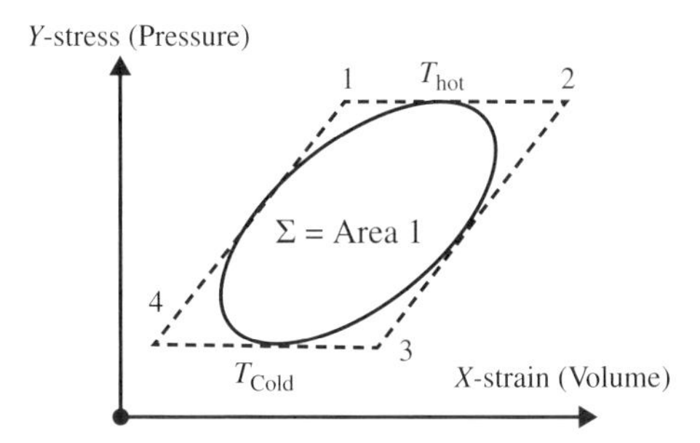

Figure 3.3 Carnot cycle in P, V plane

Both cyclic processes in Figure 3.3 operate between two temperature reservoirs: the hotter reservoir of the heat source at temperature T_{hot}; and the colder heat sink at temperature T_{cold}. We have displayed it in the figure for stress–pressure and strain–volume. However, other mechanical variables can possibly be involved besides P and V, as given in Table 1.1.

The cycle consists of four quasistatic operations: (1) an isothermal expansion (i.e., constant temperature) from 1 to 2 at temperature T_{hot} withdrawing heat δQ_{in} from the source and doing work δW_{in} (not necessarily equal to δQ_{in}), where the volume expands and the pressure decreases like a piston in a car; (2) an adiabatic expansion (i.e., no heat enters or leaves) from 2 to 3, doing further work δW_{23} but with no change in heat, ending up at temperature T_{cold}; and (3) isothermal compression at T_{cold} from 3 to 4 requiring work $-\delta W_{34} = \delta W_{43}$ to be done on the system and contributing heat $-\delta Q_{34} = \delta Q_{43}$ to the heat sink at temperature T_{cold}, ending at state 4. (4) Process 4 to 1 can be an adiabatic compression requiring work $-\delta W_{41} = \delta W_{14}$ ($\delta Q_{41} = 0$) to be done on the system to bring it back to state 1, ready for another cycle (Figure 3.3). This is a specialized sort of cycle but it is a natural one to study and one that in principle should be fairly efficient. Since the assumed heat source is at constant temperature part of the cycle must be isothermal, and if we must "dump" heat at a lower temperature we might as well give all to the lowest temperature reservoir we can find. The changes in temperature are done adiabatically.

This cycle of course does not convert all the heat withdrawn from the reservoir at T_{hot} into work; some of it is dumped as unused heat into the sink at T_{cold}. That is, more heat goes into the system than comes out:

$$Q_{in} - Q_{out} = \int_{hot} \delta Q + \int_{cold} \delta Q > 0. \tag{3.8}$$

For quasistatic processes, it is possible to decompose the energy change into the work plus heat differential form:

$$dU = \delta W + \delta Q.$$

When we do work on a specific path $P(\text{initial} \rightarrow \text{final})$ going from an initial point to a final point, then one may write

$$\begin{aligned} \Delta U &= \int_{P(\text{initial}\rightarrow\text{final})} dU = U_{final} - U_{initial}, \\ &= \int_{P(\text{initial}\rightarrow\text{final})} \delta W + \int_{P(\text{initial}\rightarrow\text{final})} \delta Q, \\ &= \delta W[P(\text{initial} \rightarrow \text{final})] + \delta Q[P(\text{initial} \rightarrow \text{final})]. \end{aligned} \tag{3.9}$$

Recall that we use the symbols δW and δQ as they are path dependent in exact differentials. The energy difference between the beginning and end of any process may simply be written as the difference $\Delta U = U_{final} - U_{initial}$, no matter the process between initial and final points. For a quasistatic process, it is possible to decompose an energy change into heat and work with the caveat that both the work $\delta W[P(\text{initial} \rightarrow \text{final})]$ and the heat $\delta Q[P(\text{initial} \rightarrow \text{final})]$ depend on the detailed nature of the path. For a simple system, the path $P(\text{initial} \rightarrow \text{final})$ may be represented as a moving point $(X(t), Y(t))$, during the time interval $t_{\text{initial}} < t < t_{\text{final}}$ of the process. For a cyclic process (closed path), one may write

$$\begin{gathered} \oint dU = \oint \delta W + \oint \delta Q = 0, \\ -\oint \delta W = \oint \delta Q. \end{gathered} \tag{3.10}$$

Here we have a closed system (Equation (3.10)) undergoing a cycle, so the internal energy ΔU is zero. During a quasistatic cycle, the work done *by* the system *on* the environment is equal to the total heat flow from the environment into the system. The net work done by the engine per cycle is the area inside 1234 in Figure 3.3, which is equal to:

$$\begin{aligned} Q_{\text{in}} - Q_{\text{out}} &= \oint \delta Q = -\oint \delta W, \\ Q_{\text{in}} - Q_{\text{out}} &= W_{\text{out}}. \end{aligned} \tag{3.11}$$

The efficiency of any heat engine may be defined as the ratio of the cyclic work output to the cyclic heat input, that is:

$$\eta_{\text{heat engine}} = \frac{\text{Network out}}{\text{Total heat in}} = \frac{W_{\text{out}}}{Q_{\text{in}}} = 1 - \left(\frac{Q_{\text{out}}}{Q_{\text{in}}}\right). \tag{3.12}$$

Now traversing the path in Figure 3.3 in a clockwise manner, the heat is moving into the system on the top part of the cycle and out of the system on the bottom part of the cycle. We therefore find the inequalities

$$Q_{\text{in}} \leq T_{\max} \Delta S \quad \text{and} \quad Q_{\text{out}} \geq T_{\min} \Delta S \tag{3.13}$$

and

$$\left(\frac{Q_{\text{out}}}{Q_{\text{in}}}\right) \geq \left(\frac{T_{\min} \Delta S_{\text{out}}}{T_{\max} \Delta S_{\text{in}}}\right) = \left(\frac{T_{\min}}{T_{\max}}\right). \tag{3.14}$$

The heat moves from the environment into the system when $\Delta S > 0$ and heat moves from the system to the environment when $\Delta S < 0$. However, the change in entropies ΔS_{in} at the hot reservoir is the same as that flowing out ΔS_{out} at the cold reservoir. From this relation we see that Equation (3.12) can be written as an inequality:

$$\eta \leq 1 - \left(\frac{T_{\min}}{T_{\max}}\right). \tag{3.15}$$

The equality yields the famous Carnot cycle efficiency:

$$\eta_{\text{Carnot}} = 1 - \left(\frac{T_{\text{min}}}{T_{\text{max}}}\right). \tag{3.16}$$

We see that the efficiency of a Carnot cycle is independent of the engine and only depends on the reservoir temperature. Therefore this is totally idealized. One way of stating the second law is to say that all Carnot cycles operating between T_{hot} and T_{cold} have the same efficiencies.

If the cycle went in the counterclockwise direction, then more heat would flow out of the system than would flow into the system. For such a cycle there is net work done on the system by the environment $W_{\text{in}} = Q_{\text{out}} - Q_{\text{in}}$. The system is then run as a refrigerator with efficiency $\eta = (W_{\text{in}}/Q_{\text{out}})$. For fixed values of the maximum and minimum temperatures during a cycle, the theorem holds true for the refrigerator efficiency as well as the engine efficiency.

3.4 Example 3.1: Cyclic Engine Damage Quantified Using Efficiency

For real engines the hot section of the Carnot cycle is from 1 to 2 where heat enters into say a "working fluid," and is then released in the Cold reservoir from 3 to 4. The temperature of the reservoir and that of the working fluid are not the same, so the process is not quite isothermal. The temperature of the hot reservoir fluid is a little cooler and the temperature of the cold reservoir fluid is a little hotter. That difference drives the efficiencies down and less work is done. For example, if the hot reservoir was at 400 K and the cold reservoir was at 200 K, the Carnot efficiency is

$$\eta_{\text{Carnot}} = 1 - \left(\frac{200\,\text{K}}{400\,\text{K}}\right) = 0.5.$$

While the theoretical exercise of a quasistatic engine was not idealized, we can think of it as a Carnot-like engine with damage. The damage creates some inefficiency where the temperature of the fluid at the 400 K reservoir is say at 390 K and the temperature of the fluid at the 200 K cold reservoir, due to some inefficiencies, is 210 K, so that the efficiency is:

$$\eta_{\text{cycle-damage}} = 1 - \left(\frac{T_{\text{min-effective}}}{T_{\text{max-effective}}}\right) = 1 - \left(\frac{210\,\text{K}}{390\,\text{K}}\right) = 0.46 < \eta_{\text{Carnot}}. \tag{3.17}$$

The actual work is of course less so the cyclic area outlined within 1234 in Area 2 of Figure 3.4 is smaller than that of the Carnot engine 1234 shown in Area 1.

One way to track cyclic engine damage is by loss of efficiency. That is, $\eta(t)$ is time dependent as the engine degrades due to cyclic damage.

For a heat engine, the heat transferred into work is less and work degrades due to engine damage increase so the work out decreases with time:

$$-\delta W_{\text{out}}(t) = \delta Q(t) = (Q_{\text{in}}(t) - Q_{\text{out}}(t)). \tag{3.18}$$

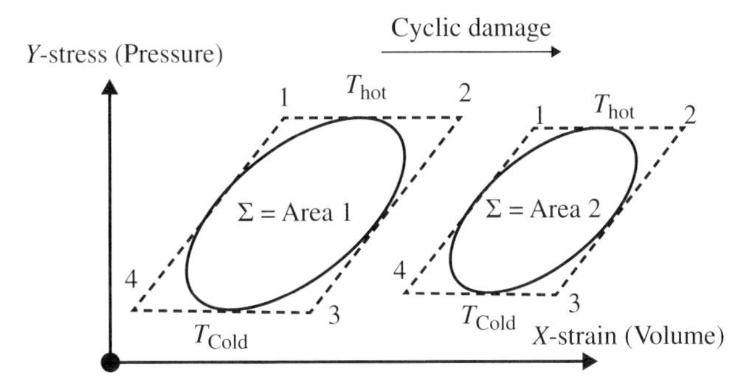

Figure 3.4 Cyclic engine damage Area 1 > Area 2

If we compare it to a new engine, for example

$$\delta Q_{\text{new-engine}} > \delta Q_{\text{degraded-engine}} \tag{3.19}$$

and

$$\eta(t) = \frac{W_{\text{out}}(t)}{Q_{\text{hot}}} = \frac{(Q_{\text{in}}(t) - Q_{\text{out}}(t))}{Q_{\text{hot}}}, \tag{3.20}$$

the actual efficiency can degrade over time. The best way to quantify the loss of efficiency for any engine, not just a heat engine, is by assessing the areas in Figure 3.4. We can assess the efficiency relative to itself when the engine was new as

$$\eta(t) = \frac{W_{\text{degraded-engine}}(\text{Area 2})}{W_{\text{new-engine}}(\text{Area 1})} = \frac{\oint_{\text{Area2}} YdX}{\oint_{\text{Area1}} YdX}. \tag{3.21}$$

When we assess damage using this method, we must keep in mind that it is tracked over the same path as work is path dependent. If for example you were to track the efficiency of your car compared to when it was new, you could assess that by noting your miles per gallon when your car was new and compare it to when your car was older, say at 100 000 miles. To do this in a reliable way, you would need the same driving route and the same type of gas. For example, you cannot compare less-efficient city driving miles to highway miles. Further, higher-octane fuels sometimes improve engine miles per gallon (MPG) performance.

Furthermore, as we noted earlier for efficiencies in Equation (2.88), anything that adds to the irreversibility of the engine such as damage reduces our efficiency:

$$\eta = \frac{W_{\text{actual}}}{W_{\text{actual}} + W_{\text{irr}}}. \tag{3.22}$$

3.5 The Thermodynamic Damage Ratio Method for Tracking Degradation

In Section 2.10.4, we provided an expression for inefficiency (Equation (2.88)); this discussion was extended in Section 3.1.1. We found the reversible work is the maximum useful work that can be obtained as a system undergoes a process between two specified states as related to the free energy. The irreversibility is lost work, which is wasted work potential during the process as a result of irreversibilities. A natural use of Equation (3.3) is in assessing damage; the thermodynamic cumulative damage can be written from the sum of the inefficiencies:

$$1-\eta=\frac{\delta W_{\text{irr}}(t)}{W_{\text{rev}}} \quad \Rightarrow \quad \text{Damage}=\frac{\sum W_{\text{actual}}(t)}{W_{\text{actual-failure}}}. \tag{3.23}$$

Here we have summed the inefficiency, used Equation (3.5) and let the maximum work to failure be equal to W_{rev}.

This is an important deduction because it says that the damage due to the irreversibility, which is hard to measure, can be assessed from the actual work, which is easier to measure.

In terms of thermodynamic degradation in cyclic work, if we are creating damage on each cycle in the system, irreversible work damage cumulates until failure occurs. Some of the irreversibility in the total process is due to inefficiencies unrelated to system work damage. This puts a bit of reality into the difficulty in assessing the true damage. In theory we should be able to track the true system work damage as it occurs, but we should distinguish the types of damage that can be measured and/or modeled. From the above equation we will note that these inefficiencies unrelated to system work damage cancel out in the actual work assessment, so that

$$\text{Damage}=\frac{\sum_{i=1}^{m} w_{\text{damage, i}}}{W_{\text{damage}}} \quad \text{and} \quad \text{Measurable Damage}=\frac{\sum_{i=1}^{m} w_{\text{total, i}}}{W_{\text{total}}}. \tag{3.24}$$

The measurable work damage ratio consists of the ratio of the ratio of the actual work performed to the actual work needed to cause system failure. In system failure, we exhaust the maximum amount of useful system work. To consistently find this damage ratio, all work found must be taken over the same work path.

The path is important as there are many ways to walk up a hill; therefore, we will always require the numerator and denominator to travel the same path. For the damage expression, $w_{\text{damage},i}$ is the cyclic work damage performed over i cycles, and W_{damage} is the total work damage performed to cause system failure. This damage is created in the system by the environment doing work on the system, or *vice versa*.

In this expression we write W_{total} as the total actual work performed to failure and $w_{\text{total},i}$ as the total actual cyclic work performed over i cycles. As a mechanical work example, damage might be related to plastic strain; we usually can measure and/or model this.

We can now define cyclic damage over n cycles using the conjugate work variables in Table 1.1 [1, 2] as:

$$\text{Cyclic damage} = \frac{\sum_{n} \oint Y_n dX_n}{\sum_{n=1}^{N} \oint Y_n dX_n} = \frac{\sum_{n} \oint Y_n dX_n}{W_{\text{failure}}} \tag{3.25}$$

where N represents the number of cycles to failure and n represents the number of cycles prior to failure. When the damage ratio value is 1, failure results for which we require the total work summed along the path traveled in the numerator to equal the full work to failure that occurred along the same path traveled in the denominator.

In many cases it is actually possible to sum the work for each cycle in a consistent measurement process. For example, if Y is stress and X is plastic strain, we can use strain gages and monitor the stress level and integrate out the area after each cycle with computer software. An accurate estimate can therefore be made of the damage occurring. In other cases where such measurements are not possible, often what is used is the Miner's Rule [3] that we found in Section 2.1.1. This often serves as a good approximation of the effective damage and is detailed further in Section 4.2.

If the work is non-cyclic with only one kind of work involved, and we are careful to ensure that the path traveled in the numerator is identical to that in the denominator, than the thermodynamic damage is [1, 2]:

$$\text{Damage} = \frac{\int Y dX}{W_{\text{failure}}} = \frac{\text{Partial work}}{\text{Total work needed for failure}}. \tag{3.26}$$

If there are other types of work stress causing damage, then these can be summed as long as the work path is the same. That is, we travel up the hill on the same path but we may at times carry a different amount of weight along the way. If we travel the path with one weight we may break down just as we reach the top of the hill, while the next time we go up the path with more weight, we will break down sooner. The key is the path traveled, and then we can consider different stresses as we travel this path. Then we should be able to assess the second time up the hill if and when we will break down relative to the breakdown value of 1. In the case of other stress, we write

$$\text{Work} = \sum_{i} \int Y_i \, dX_i. \tag{3.27}$$

This is our approach for assessing non-equilibrium thermodynamic damage. Many examples are provided in Chapter 4.

3.6 Acceleration Factors from the Damage Ratio Principle

Reliability testing utilizes the notion of time compression to perform, say, 5 years of testing in a practical experimental time frame. The acceleration factor (AF) helps us to estimate this time compression. For example, if 1 month of testing represents 1 year in the field, then the AF is 12. AFs can be derived from first principles utilizing the thermodynamic work and the damage concept. Here we explain how this is done and, in Chapter 4, we derive from first principles numerous AFs in mechanical systems used in reliability testing. See also temperature AF in

Chapter 5 and the diffusion AF in Chapter 8. In the Special Topics B section we provide application for their uses.

In NE thermodynamics, as we seek to trace the degradation process over time Equation (3.6) will often have the separable form

$$w=\int_{t_{\text{initial}}}^{t_{\text{final}}} Y(t)\frac{dX(t)}{dt}dt=f(Y,k,E_{\text{a}})t \tag{3.28}$$

where t is time, $f(Y, k, E_{\text{a}})$ is some function of the environmental stress, and k and E_{a} are specific constants related to the degradation mechanism such as a specific power exponent and activation energy. Then according to Equations (3.25) and (3.26), if we have damage between two different environmental stresses Y_1 and Y_2, and failure occurs for each at time τ_1 and τ_2 respectively, then the damage value equal to 1 requires that [4]:

$$\text{Damage}=\frac{f(Y_2,k,E_{\text{a}})\,\tau_2}{f(Y_1,k,E_{\text{a}})\,\tau_1}=1 \text{ or } \text{AF}_{\text{damage}}(1,2)=\frac{\tau_2}{\tau_1}=\frac{f(Y_1,k,E_{\text{a}})}{f(Y_2,k,E_{\text{a}})} \tag{3.29}$$

where $\text{AF}_{\text{damage}}(1,2)$ is an AF used in reliability testing, often called the time compression AF between the two different stress environments 1 and 2. As an introduction to the reader who is not familiar with accelerated testing, the AF is used to estimate test time as follows:

$$\text{Test time}=\frac{\text{Estimated product lifetime}}{\text{AF(test, use)}}=\text{Time compression.} \tag{3.30}$$

Multiple examples are given in Special Topics B on AF usage, and numerous AFs are found using the damage principle in the next chapter. The concept is that, by raising the level of stress, a test can simulate life conditions over a much shorter time period. The AF is then determined by the stress equation relative to use conditions. The notation AF(test, use) then refers to the AF between test and use environments. Note that, in general, accelerating time requires $\text{AF} > 1$.

Then for any ith stress condition, the time to failure τ_i when $\text{AF}_{\text{damage}}(1,i)$ is known is

$$\tau_i=\tau_1 \text{ AF}(1,i) \text{ and } \text{AF}_{\text{damage}}(1,i)=\frac{f(Y_1,k,E_{\text{a}})}{f(Y_i,k,E_{\text{a}})}. \tag{3.31}$$

When $\text{AF}_{\text{damage}}$ is known, it allows us to write the damage at any time t between two stress environments along the same work path as [4]:

$$\text{Damage}=\frac{f(Y_1,k,E_{\text{a}})\,t_1}{f(Y_2,k,E_{\text{a}})\,\tau_2}=\frac{f(Y_1,k,E_{\text{a}})\,t_1}{f(Y_2,k,E_{\text{a}})\,\text{AF}_{\text{damage}}(1,2)\,\tau_1}. \tag{3.32}$$

The damage AF is an important tool in assessing the time evolution of thermodynamic damage occurring between different environments. A number of AFs are derived in the next chapter. Examples of how to use the AF in testing is found in Special Topics B.

Summary

3.1.1 Conjugate Work and Free Energy Approach to Understanding Non-Equilibrium Thermodynamic Degradation

As entropy damage increases for the system, the system's free energy decreases and the system loses its ability to do useful work. Then instead of focusing on entropy, we can look at the free energy of a system. For example, the Helmholtz free energy is

$$\delta W = -dF + TdS_{\text{damage}} \tag{3.3}$$

for conditions that $dT = dN = 0$. This above equation can be interpreted as (see also Section 2.10.4):

$$W_{\text{actual}} = W_{\text{rev}} - W_{\text{irr}}. \tag{3.4}$$

This statement proves to be relevant to many thermodynamic work processes. It is a statement about the free energy that equals the reversible work here, the maximum useful available work of the system.

3.2 Application to Cyclic Work and Cumulative Damage

For a cyclic process C in a simple system, the work done on the system by the environment is given by the area obtained from the integral around a closed curve:

$$\text{Work(cycle)} = \oint_C Y dX = \int_{t_{\text{initial}}}^{t_{\text{final}}} Y(t) \frac{dX(t)}{dt} dt. \tag{3.6}$$

3.3 Cyclic Work Process, Heat Engines, and the Carnot Cycle

For a cyclic process (closed path), one may write:

$$\begin{gathered} \oint dU = \oint \delta W + \oint \delta Q = 0, \\ -\oint \delta W = \oint \delta Q. \end{gathered} \tag{3.10}$$

The net work done by the engine per cycle is the area inside 1234 in Figure 3.3, which is equal to

$$\begin{gathered} Q_{\text{in}} - Q_{\text{out}} = \oint \delta Q = -\oint \delta W, \\ Q_{\text{in}} - Q_{\text{out}} = W_{\text{out}}. \end{gathered} \tag{3.11}$$

The efficiency of any heat engine may be defined as the ratio of the cyclic work output to the cyclic heat input:

$$\eta_{\text{heat engine}} = \frac{\text{Net work out}}{\text{Total heat in}} = \frac{W_{out}}{Q_{in}} = 1 - \left(\frac{Q_{out}}{Q_{in}}\right). \tag{3.12}$$

From this relation and Equation (3.12), this was written as an inequality

$$\eta \leq 1 - \left(\frac{T_{min}}{T_{max}}\right). \tag{3.15}$$

The equality yields the famous Carnot cycle efficiency:

$$\eta_{\text{Carnot}} = 1 - \left(\frac{T_{min}}{T_{max}}\right). \tag{3.16}$$

We see that the efficiency of a Carnot cycle is independent of the engine and only depends on the reservoir temperature. A Carnot cycle is therefore a theoretical quasistatic operation and operates too slowly to be practical.

3.4 Example 3.1: Cyclic Engine Damage Quantified Using Efficiency

The best way to quantify the loss of efficiency for any engine, not just a heat engine, is by assessing the areas in Figure 3.4. We can assess the efficiency relative to itself when the engine was new as:

$$\eta(t) = \frac{W_{\text{degraded-engine}}(\text{Area 2})}{W_{\text{new-engine}}(\text{Area 1})} = \frac{\oint_{\text{Area 2}} YdX}{\oint_{\text{Area 1}} YdX}. \tag{3.21}$$

3.5 The Thermodynamic Damage Ratio Method for Tracking Degradation

In theory we should be able to track the true system work damage as it occurs, but we should distinguish the types of damage that can be measured and/or modeled as:

$$\text{Damage} = \frac{\sum_{i=1}^{m} w_{\text{damage, i}}}{W_{\text{damage}}} \quad \text{and} \quad \text{Effective damage} = \frac{\sum_{i=1}^{m} w_{\text{total, i}}}{W_{\text{total}}}. \tag{3.24}$$

The measurable work damage ratio consists of the ratio of the actual work performed to the actual work needed to cause system failure. In system failure, we exhaust the maximum amount of useful system work. To consistently find this damage ratio, all work found must be taken over the same work path.

We can now define cyclic damage over n cycles using the conjugate work variables in Table 1.1 as [1, 2]

$$\text{Cyclic damage} = \frac{\sum_{n} \oint Y_n dX_n}{W_{\text{failure}}}. \tag{3.25}$$

If the work is non-cyclic with only one kind of work involved, and we are careful to ensure that the path traveled in the numerator is identical to that in the denominator, than the thermodynamic damage is

$$\text{Damage} = \frac{\int YdX}{W_{\text{failure}}} = \frac{\text{Partial work}}{\text{Total work needed for failure}}. \tag{3.26}$$

3.6 Acceleration Factors from the Damage Ratio Principle

In NE thermodynamics, as we seek to trace the degradation process over time t, Equation (3.6) will often have the separable form

$$w = \int_{t_{\text{initial}}}^{t_{\text{final}}} Y(t)\frac{dX(t)}{dt}dt = f(Y,k,E_{\text{a}})t. \tag{3.28}$$

Then for any ith stress condition, the time to failure τ_i when $\text{AF}_{\text{D}}(1,i)$ is known as

$$\tau_i = \tau_1 \ \text{AF}(1,i) \ \text{ and } \ \text{AF}_{\text{damage}}(1,i) = \frac{f(Y_1,k,E_{\text{a}})}{f(Y_i,k,E_{\text{a}})}. \tag{3.31}$$

When $\text{AF}_{\text{damage}}$ is known, it allows us to write the damage at any time t between two stress environments along the same work path as

$$\text{Damage} = \frac{f(Y_1,k,E_{\text{a}})\,t_1}{f(Y_2,k,E_{\text{a}})\,\tau_2} = \frac{f(Y_1,k,E_{\text{a}})\,t_1}{f(Y_2,k,E_{\text{a}})\,\text{AF}_{\text{damage}}(1,2)\,\tau_1}. \tag{3.32}$$

The damage AF is an important tool in assessing the time evolution of thermodynamic damage occurring between different environments.

References

[1] Feinberg, A. and Widom, A. (2000) On thermodynamic reliability engineering. *IEEE Transaction on Reliability*, **49** (2), 136.
[2] Feinberg, A., Crow, D. (eds) (2001) *Design for Reliability*, M/A-COM 2000, CRC Press, Boca Raton.
[3] Miner, M.A. (1945) Cumulative damage in fatigue. *Journal of Applied Mechanics*, **12**, A159–A164.
[4] Feinberg, A. (2015) Thermodynamic damage within physics of degradation, in *The Physics of Degradation in Engineered Materials and Devices* (ed J. Swingler), Momentum Press, New York.

4

Applications of NE Thermodynamic Degradation Science to Mechanical Systems: Accelerated Test and CAST Equations, Miner's Rule, and FDS

4.1 Thermodynamic Work Approach to Physics of Failure Problems

We are now in a position to assess a number of reliability degradation problems. In this chapter we will continue to use the thermodynamic work to assess the entropy damage to the system and the amount of free energy left for numerous key problems in reliability. As we found in Chapter 3, *the thermodynamic work is perhaps the most directly measurable and practical quantity to use in assessing a system's free energy and its degradation* [1–3]. This chapter provides many examples for deriving physics of failure aging laws and their acceleration factors in mechanical systems using the thermodynamic work principle.

4.2 Example 4.1: Miner's Rule

We derived the Miner's rule using an entropy approach in Section 2.1.1. Here we derive it using the thermodynamic work approach [1–3]. Equation (3.25) for cyclic damage may be used to derive Miner's empirical rule [4], commonly used for accumulated fatigue damage and a number of other useful expressions in damage assessment of devices and machines.

Consider a system undergoing fatigue as in the paper clip example, where we bend it back and forth a certain distance for three cycles. To find the actual work we need to sum the cyclic work area of each bend cycle since both the stress σ and strain e will change

Thermodynamic Degradation Science: Physics of Failure, Accelerated Testing, Fatigue, and Reliability Applications, First Edition. Alec Feinberg.

slightly as the paper clip fatigues. If we use Stokes theorem, it demonstrates the work is related to the cyclic area of each:

$$w_1(\text{cycle}) = \sum_{i=1}^{3} \oint_{\text{Area } i} \sigma de = \iint_{\text{Area 1}} d\sigma de + \iint_{\text{Area 2}} d\sigma de + \iint_{\text{Area 3}} d\sigma de. \tag{4.1}$$

In Miner's rule an approximation is actually made. Miner empirically figured that stress σ and cycles n were the main factors of damage.

In our framework this means

$$\text{Work}_n = \text{Work}(\sigma, n).$$

Miner also empirically assumed that the work for n cycles of the same cyclic size is all that is needed. In our framework this means

$$\text{Work}_n = n \text{ Work}(\sigma)$$

(Miner's assumption versus reality, as work is reduced each σ cycles.)

Using this assumption, then we average the three cyclic areas of the same stress multiplied by three in our case:

$$w_i = 3\oint \sigma_i de. \tag{4.2}$$

Following this approximation, for any stress level we just count the number of cycles so that the total thermodynamic work is

$$\begin{aligned} W(\text{for } k \text{ types of stress}) &= n_1\oint \sigma_1 de + n_2\oint \sigma_2 de + n_3\oint \sigma_3 de + \ldots \\ &= n_1 W_1 + n_2 W_2 + n_3 W_3 + \ldots = \sum_{i=1}^{K} n_i W_i. \end{aligned} \tag{4.3}$$

Using this approximation, we obtain effective damage (i.e., not the true damage) where the cumulative effective damage is:

$$\text{Effective damage} \approx \frac{\sum_{i=1}^{K} n_i w_i}{W_{\text{failure}}} \tag{4.4}$$

or

$$\begin{aligned} \text{Effective damage} &\approx \frac{n_1\oint \sigma_1 de + n_2\oint \sigma_2 de + n_3\oint \sigma_3 de + \ldots}{W_{\text{failure}}} \\ &= \frac{n_1\oint \sigma_1 de}{W_{\text{failure}}} + \frac{n_2\oint \sigma_2 de}{W_{\text{failure}}} + \frac{n_3\oint \sigma_3 de}{W_{\text{failure}}} + \ldots \end{aligned} \tag{4.5}$$

Using the approximation above, the total work to failure is the same for each cyclic size type along the same work path so that

$$W(\text{for failure}) = N_1 W_1 = N_2 W_2 = N_3 W_3 = \ldots \tag{4.6}$$

So this yields

$$\begin{aligned}\text{Effective damage} &\approx \frac{n_1 \oint \sigma_1 de}{W_{\text{failure}}} + \frac{n_2 \oint \sigma_2 de}{W_{\text{failure}}} + \frac{n_3 \oint \sigma_3 de}{W_{\text{failure}}} + \ldots \\ &= \frac{n_1 \oint \sigma_1 de}{N_1 W_1} + \frac{n_2 \oint \sigma_2 de}{N_2 W_2} + \frac{n_3 \oint \sigma_3 de}{N_3 W_3} + \ldots\end{aligned} \tag{4.7}$$

giving

$$\begin{aligned}\text{Effective damage} &\approx \frac{n_1 \oint \sigma_1 de}{N_1 \oint \sigma_1 de} + \frac{n_2 \oint \sigma_2 de}{N_2 \oint \sigma_2 de} + \frac{n_3 \oint \sigma_3 de}{N_3 \oint \sigma_3 de} + \\ &= \frac{n_1}{N_1} + \frac{n_2}{N_2} + \frac{n_3}{N_3} + \ldots\end{aligned} \tag{4.8}$$

Therefore, Miner's rule is an approximation of the cumulative fatigue damage, commonly written

$$\text{Effective damage} \approx \frac{n_1}{N_1} + \frac{n_2}{N_2} + \frac{n_3}{N_3} + \ldots = \sum_{i=1}^{k} \frac{n_i}{N_i}. \tag{4.9}$$

An example of how to use Miner's rule is given in Section 2.1.2. This is in agreement with what we found in Section 2.1.1 using the entropy approach. Equation (4.4) (also see Equation (3.25)) is a better estimate of the damage, as Equation (4.9) is based on the assumption that the stress is independent of the number of cycles tested. In fact we know from experience that, as the paper clip fatigues, it takes less stress to create the same bend amplitude.

4.2.1 Acceleration Factor Modification of Miner's Damage Rule

It can be difficult to always know the cycles to failure at any *i*th stress level in Miner's application. However, since we are on the same work path, we can often establish a cyclic acceleration factor (see Special Topic B for examples) between stress levels. If AF_{damage} and N_1 are known, then Miner's rule can be simplified from Equation (4.9) for the *i*th stress using

$$N_i = AF_{\text{damage}}(1, i)\, N_1. \tag{4.10}$$

Then Miner's rule can be modified as [3, 4]:

$$\text{Effective damage} \approx \frac{n_1}{N_1} + \frac{n_2}{\text{AF}_{\text{damage}}(1,2)N_1} + \frac{n_3}{\text{AF}_{\text{damage}}(1,3)N_1} + \ldots$$
$$= \sum_{i=1}^{k} \frac{n_i}{\text{AF}_{\text{damage}}(1,i)N_1}. \tag{4.11}$$

The acceleration factor $\text{AF}_{\text{damage}}$ is given in Section 4.3.4 and depends on the type of vibration (sine or random). See also Section 4.5 for use of this equation in fatigue damage spectrum (FDS) analysis. It can also be used in thermal cycling (see Section 4.3.3).

4.3 Assessing Thermodynamic Damage in Mechanical Systems

In this section we will provide examples of how to apply the concept of damage to a number of mechanical systems of creep and wear, as well as an analysis for cyclic vibration and thermal fatigue.

4.3.1 Example 4.2: Creep Cumulative Damage and Acceleration Factors

Creep parameters include the strain (ε) length change ΔL ($\Delta L/L$) due to an applied stress (σ) at temperature (T). In the elastic region, stress causes a strain that is recoverable (i.e., reversible) so that $\sigma = Y\varepsilon$ where Y is Young's modulus. When stress increases such that the work is irreversible, the permanent plastic strain ε_p damage is termed plastic deformation. The most popular empirical creep rate equation is [3, 5]

$$\varepsilon_p = B_0 \sigma^M t^p e^{-E_a/k_B T} \tag{4.12}$$

where B_0 is a material strength constant; p is the time exponent where $0 < p < 1$ for primary and secondary creep stages; M (where $0 < 1/M < 1$) is strain hardening exponent, dependent on the material type; and E_a is the thermal activation energy for the creep process (see Figure 9.5 for the three stages of creep). We note a linear time dependence is observed in the secondary creep phase where $p = 1$. Some tabulated values are given in Table 4.1 for M and B_0 for the secondary creep rate.

Table 4.1 Typical constants for stress–time creep law

Material	Temperature (°C)	B_0 (in^2/lb)^N per day	M
1030 Steel	400	48×10^{-38}	6.9
1040 Steel	400	16×10^{-46}	8.6
2Ni-0.8Cr-0.4Mo Steel	454	10×10^{-20}	3.0
12Cr Steel	454	10×10^{-27}	4.4
12Cr-3W-0.4Mn Steel	550	15×10^{-16}	1.9

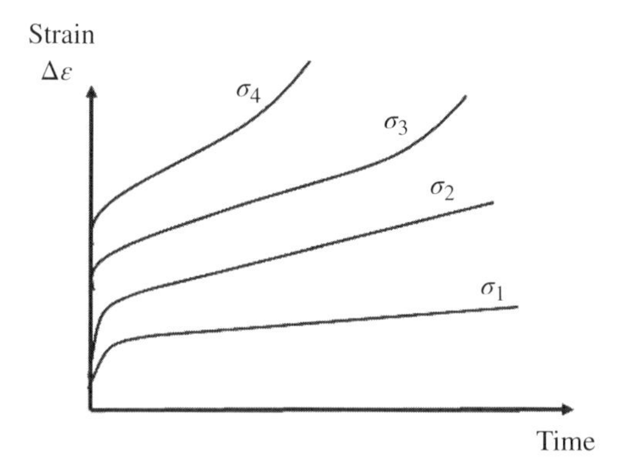

Figure 4.1 Creep strain over time for different stresses where $\sigma_4 > \sigma_3 > \sigma_2 > \sigma_1$

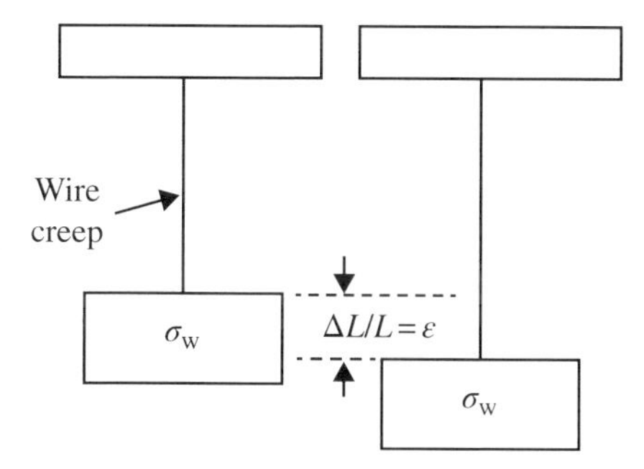

Figure 4.2 Example of creep of a wire due to a stress weight

The thermodynamic work causing damage from the creep process of the metal is found from the stress–strain creep area when the stress–strain relation is plotted in Figure 4.1 to demonstrate typical creep data.

The conjugate work variables for creep are stress σ for force and $\varepsilon = \Delta L/L$ strain, the length variable. From Figure 4.1 displaying creep, we can see that it is logical to model the strain over time by a power law; the strain will also be a power law as a function of stress. We know from Equation (4.12) that these are good assumptions (see Figure 4.1), so that our general expression is

$$\varepsilon = A\sigma^b t^Y. \tag{4.13}$$

Furthermore, from Figure 4.1 it is logical that $0 < M < 1$ for primary–secondary creep region (Figure 4.2).

The thermodynamic work causing damage from the creep process depicted in Figure 4.2 is found from the stress–strain creep area when the stress–strain relation is plotted. Assessing the damage is more accurately found if the data were available. Here, we can use the empirical creep rate expression to find this area by integrating the expression above to determine the work in terms of the applied stress, which is more easily known. The creep path is dL ($\varepsilon = \Delta L/L$) for the integral limits, so summing the thermodynamic work we have

$$w=\int_0^{\Delta L}\sigma d\varepsilon=\int_0^{\Delta L}\sigma\frac{d\varepsilon}{dt}dt=A\sigma^{b+1}t^Y=A(T)\bar{\sigma}^M t^p. \tag{4.14}$$

Making comparisons to the original empirical creep equation (Equation (4.12)), we have $A(T)$ is defined within a constant $B=B_0\exp(-Q/RT)$, while $M=b+1$ and $Y=p$. We have assumed that stress is not time dependent. Note that we prefer to use the term $d\varepsilon/dt$ for a constant stress to integrate over time to obtain damage in terms of the applied stress.

In order to assess the damage, we need to have some knowledge of the critical damage at a particular stress and temperature. Let's assume this occurs at time τ_1 at stress level σ_1 and temperature T_1. Then the thermodynamic damage ratio at any other stress σ_2 and temperature T_2 at time t_2 along the same work path is

$$\text{Creep damage}=\frac{w}{W_{\text{failure}}}=e^{-Q/k_B\left(\frac{1}{T_2}-\frac{1}{T_1}\right)}\left(\frac{\sigma_2}{\sigma_1}\right)^M\left(\frac{t_2}{\tau_1}\right)^p \tag{4.15}$$

where $W_{\text{failure}}=A(T_1)\sigma_1^M\tau_1^p$ and $w=A(T_2)\sigma_2^M t_2^p$. Here, τ_i is time to failure (a constant) at stress i and t_i is work time at stress level i. Again, this is valid for stresses 1, 2 along the same work path ΔL. If damage is 1, failure occurs (at $t_2=\tau_2$) and we can then write this as

$$\text{AF}_{\text{creep}}=\left(\frac{\tau_2}{\tau_1}\right)=e^{-E_a/k_B\left(\frac{1}{T_1}-\frac{1}{T_2}\right)}\left(\frac{\sigma_1}{\sigma_2}\right)^K. \tag{4.16}$$

This is the creep acceleration factor (see Special Topic B on Acceleration Factor usage). Recall that $W_{\text{failure1}}=W_{\text{failure2}}$ (i.e., work to failure stress 2 and 1; see Equation (4.6)), so their ratio will be 1. Here $M/p=K$ and $Q=E_a/p$. Often it is helpful to write the linear form for time to failure for a particular stress. This is deduced as follows from the above equation

$$\tau_2=\tau_1 e^{-E_a/k_BT_1}(\sigma_1)^K\left(\frac{1}{\sigma_2}\right)^K e^{E_a/k_BT_2}=C\left(\frac{1}{\sigma_2}\right)^K e^{E_a/k_BT_2}. \tag{4.17}$$

Writing the time to failure t_{failure} for τ_2 we have

$$\ln(t_{\text{failure}})=C+\frac{E_a}{k_BT}-K\ln(\sigma) \tag{4.18}$$

where the constant C can always be written in terms of another stress if we know that $C=\tau_1 e^{-E_a/k_BT_1}(\sigma_1)^K$; however, it also equates to the factored-out constant in Equation (4.15) $C=\ln(1/A)$.

Lastly, it should be noted that if a number of different i stresses are applied, the general creep damage ratio can be written by accumulating the thermodynamic damage along the same work path (Equation (4.15)) as

$$\text{Damage} = \sum_i \left(\frac{t_i}{\tau_i}\right)^p = \sum_i \left(\frac{t_i}{\text{AF}_{\text{creep}}(1,i)\tau_1}\right)^p. \tag{4.19}$$

In Section 4.5 and Special Topics B.7 we illustrate how this equation can be used for environmental profiling in order to find a cumulative accelerated stress test (CAST) goal.

4.3.2 *Example 4.3: Wear Cumulative Damage and Acceleration Factors*

Wear nicely exemplifies the thermodynamic approach. There are many different types of wear, including abrasive, adhesive, fretting, and fatigue wear. The most common wear model used for adhesive and sometimes abrasive wear of the softer material between two sliding surfaces is the Archard's wear equation [6, 7]:

$$D = \frac{kP_W l}{AH} \tag{4.20}$$

where D is removed depth of the softer material; P_W is normal load (lb); l is the sliding distance (feet); H is hardness of the softer material (psi); A is contact area; and k is Archard's wear coefficient (dimensionless). In the adhesive wear of metals [6–8], wear coefficient k varies between 10^{-7} and 10^{-2} depending on the operating conditions and material properties. It should be recognized that a wear coefficient k is constant typically only within a certain adhesive wear-rate range.

We define this as our system as shown in Figure 4.3, consisting of the weight, contact area, hardness, and k value and which is traveling at some velocity. Next we need to provide the environment which is responsible for doing work on our system which causes the movement so that wear can occur. We introduce an external force P_E as shown in Figure 4.3.

P_F creates the sliding and induces the thermodynamic wear work. Some of the external work goes into creating the wear velocity while some causes wear. Work is simply the integral of the force times distance or $w = \int P_F dx$ and dx is the sliding distance. It is logical that P_F is proportional to P_W, but since we are causing wear we write it as $P_F = C P_W$. The distance dx can be written as $dx = C_2 dL/dt\, dt = C_2 v dt$, where C_2 is a constant related to the surface wear friction

$$w = \int_{x_1}^{x_2} P_F dx = \int_{t_1}^{t_2} (CP_W) C_2 \frac{dL}{dt} dt = CC_2 P_W vt = k_P \frac{kP_W \nu t}{AH}. \tag{4.21}$$

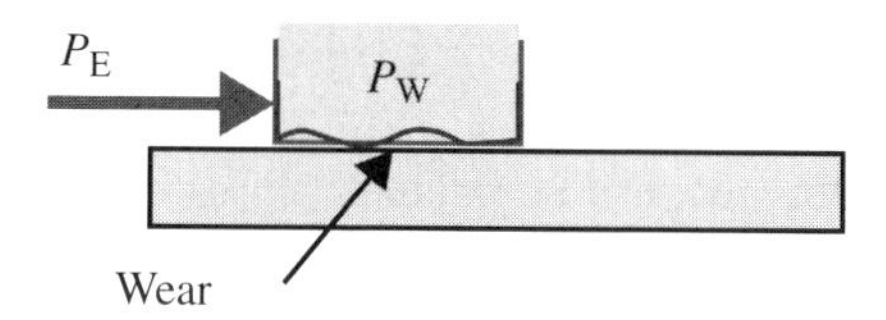

Figure 4.3 Wear occurring to a sliding block having weight P_W

Note that the integral is over x_1 to x_2, the work path in the time t_1 to t_2, which is linear with wear amount. Since we have substituted vt for the sliding distance, we need to keep this in mind. Here we have introduced the knowledge of Archard's wear model by identifying that our constant C_2 is k/AH times some constant $k_P = C$ that is related to the work type. This constant is needed as without it we would just have Archard's equation which has units of distance not work.

At some time $t = \tau_1$ we may consider that too much damage has occurred in a stress environment we call 1 as compared to stress environment 2. Then we can assess the damage ratio between environments 1 and 2, where 1 has caused failure and in 2 failure has not yet occurred:

$$\text{Wear damage} = \frac{w_2}{W_1} = \frac{\left(k_P \dfrac{P_{W2}\nu_2 t}{HA_2}\right)}{\left(k_P \dfrac{P_{W1}\nu_1 \tau_1}{HA_1}\right)} = \frac{\left(P_{W2}\dfrac{\nu_2 t}{A_2}\right)}{\left(P_{W1}\dfrac{\nu_1 \tau_1}{A_1}\right)} \tag{4.22}$$

where τ_1 is the time for critical wear failure in stress environment 1. Note that for the two environments (1 and 2) the materials are the same so H and k cancel out. We must also stay on the same work type to make comparisons between environments; therefore k_P are the same.

When failure occurs (the amount of wear is the same for both environments), the thermodynamic work in environment 2 equals that of environment 1, the damage ratio is 1, and $W_{F1} = W_{F2}$. This occurs at time τ_2 in environment 2, so we can then write the wear acceleration factor as

$$\text{AF}_{\text{wear}}(2,1) = \frac{\tau_1}{\tau_2} = \left[\frac{\left(P_{W2}\dfrac{\nu_2}{A_2}\right)}{\left(P_{W1}\dfrac{\nu_1}{A_1}\right)}\right]_{\text{DF}}. \tag{4.23}$$

(See Special Topics B for how to apply acceleration factors in testing.) The subscript DF indicates that we have created the same wear amount causing failure in both environments. One thing we note is that Archard's type of wear is not easily accelerated as it is linear with weight and velocity.

It is helpful to write the linear form for time to failure for a particular stress. This is deduced to within a constant from the above equation

$$\tau_1 = \tau_2 P_{W2}\frac{\nu_2}{A_2}\left[\frac{1}{\left(P_{W2}\dfrac{\nu_1}{A_1}\right)}\right] = C\left(\frac{A_1}{P_{W1}\nu_1}\right). \tag{4.24}$$

Then the time to failure for wear can be written $t_{\text{failure}} = \tau$:

$$t_{\text{failure}} = C\frac{A}{P_{W1}\nu_1} \tag{4.25}$$

where C is a constant that can be written in terms of another stress level if known. However, it also represents the ratio of the factored-out constant $C = H/k_p k$.

If a number of different i stresses are applied along the same work path (Equation (4.22)). The general wear damage ratio can be written by accumulating the thermodynamic damage at each stress level as

$$\text{Damage} = \sum_i \left(\frac{t_i}{\tau_i}\right) = \sum_i \left(\frac{t_i}{\text{AF}_{\text{wear}}(1,i)\tau_1}\right). \tag{4.26}$$

In Section 4.3 and Special Topic B.7 we illustrate how this equation can be used for environmental profiling in order to find a CAST goal.

If the external force causes oscillator motion so that we are doing cyclic work for n cycles, we can replace t by n and τ by N to obtain Miner's Rule for Archard's type wear.

4.3.3 *Example 4.4: Thermal Cycle Fatigue and Acceleration Factors*

In thermal cycling, a temperature change ΔT in the environment from one extreme to another causes expansion and/or contraction (i.e., strain) in a material system. The plastic strain (ε) caused by the thermal cyclic stress (σ) in the material can be written similar to Equation (4.12) as:

$$\varepsilon_P = \beta_0 n^Y \Delta T^b e^{-E_a/k_B T} \tag{4.27}$$

where we have substituted $\sigma^M \sim \Delta T^b$ for the non-linear stress and n for thermal cycles instead of time.

The thermodynamic work causing damage from thermal cycle stress is found from the stress–strain creep area if the stress–strain relation could be plotted. Then, similar to creep work above, the cyclic work is still over the path ΔL to Equation (4.14). However, the work path is not much of an issue in say solder joint expansion contraction or other joints that may have different expansion–contraction rates. The main concern is keeping the system, such as the solder joint, consistent during analysis. We have

$$w = \oint_{\Delta L} \sigma d\varepsilon = \oint_{\Delta L} \Delta T \frac{d\varepsilon}{dn} dn = B(T)\Delta T^{b+1} n^Y = B(T)\Delta T^M n^p \tag{4.28}$$

where $B(T) = \beta_0 e^{-E_a/k_B T}$. We need to have some knowledge of the critical damage at a particular cyclic stress and along this work path. Let's assume this occurs at N_1 at stress level ΔT_1. Then, similar to Equation (4.15), the damage ratio at another stress ΔT_2 for n_2 cycles is

$$\text{Thermal cycle damage} = e^{-E_a/k_B\left(\frac{1}{T_2} - \frac{1}{T_1}\right)} \left(\frac{\Delta T_2}{\Delta T_1}\right)^M \left(\frac{n_2}{N_1}\right)^p. \tag{4.29}$$

If damage is 1 then $n_2 = N_2$, failure occurs and we write the acceleration factor

$$AF_{\text{cyclic fatigue}} = \left(\frac{N_2}{N_1}\right) = e^{-E_a/k_B\left(\frac{1}{T_1}-\frac{1}{T_2}\right)}\left(\frac{\Delta T_1}{\Delta T_2}\right)^K \tag{4.30}$$

where $k/p = K$ and $Q = E_a/p$. The non-Arrhenius ratio is called the "Coffin-Manson" acceleration factors [9–11], or Equation (4.30) is the Modified Coffin-Manson acceleration factor. When the activation energy E_a is small then the Arrhenius effect can be neglected. For example, in solder joint testing, $K = 1.9$ for lead-free solder and about 2.5 for lead solder. The activation energy is about 0.123 eV that is typically used. For example, if use condition is stress level 1 cycle between 20 and 60 ($\Delta T = 40°C$, $T_{max} = 60°C$), while test stress condition is stress level 2 cycled between −20 and 100°C ($\Delta T = 120°C$, $T_{max} = 100°C$), then Arrhenius AF = 1.58 while the Coffin-Manson AF = 9 with an overall AF of 14.2. In the case where we have 1 cycle per day in use condition, we see that 10 years of use condition is about 260 test cycles (see Special Topics B for more applications).

Equation (4.30) is also similar to the Norris–Lanzberg [12] thermal cycle model which also includes a thermal cycle frequency effect (see Equation (B18) for details and uses of their model).

It is helpful to write the linear form for cycles to failure for a particular stress. This is deduced as

$$N_2 = N_1 e^{-E_a/k_B T_1}(\Delta T_1)^{-K}\left(\frac{1}{\Delta T_2}\right)^K e^{E_a/k_B T_2} = C\left(\frac{1}{\Delta T_2}\right)^K e^{E_a/k_B T_2}. \tag{4.31}$$

The thermal cycles to failure is therefore

$$\ln(N_{\text{failure}}) = C + E_a/Tk_B - K\ln(\Delta T) \tag{4.32}$$

where C can be written in terms of the other stress factor if known, $C = N_1 e^{-E_a/k_B T_1}(\Delta T_1)^{-K}$. However, it can also be found as the factored-out ratio constant $C = \ln(1/B)$.

Lastly, it should be noted that if a number of different stresses are applied the general damage ratio can be written by accumulating the thermodynamic damage along the same work path (Equation (4.29)) as in Miner's rule:

$$\text{Damage} = \sum_i \left(\frac{n_i}{N_i}\right)^p = \sum_i \left(\frac{n_i}{AF_{\text{cyclic fatigue}}(1,i)N_1}\right)^p. \tag{4.33}$$

In Section 4.4 and Special Topics B.7 we illustrate how this equation can be used for environmental profiling in order to find a CAST goal.

4.3.4 *Example 4.5: Mechanical Cycle Vibration Fatigue and Acceleration Factors*

In a similar manner to the above argument for thermal cycle, we can find the equivalent mechanical vibration cyclic fatigue damage. In a vibration environment, a vibration level depends on the type of exposure. In testing two types of environments are typically used: sinusoidal and random vibration profiles. In sinusoidal vibration the stress level is denoted G_s, where G is a unitless quantity equal to the sinusoidal acceleration A divided by the gravitational constant g. In random vibration, a similar quantity is used termed G_{rms} (defined below). Consider first the plastic strain (ε) caused by a sinusoidal vibration level G stress (σ) in the material. The strain can be written similarly to Equation (4.27):

$$\varepsilon = \beta_0 n^p G^j. \tag{4.34}$$

The cyclic work is then found, similar to Equation (4.28), as

$$w = \oint_{\Delta L} \sigma d\varepsilon = \oint_{\Delta L} G \frac{d\varepsilon}{dn} dn = A G^{j+1} n^P = G^Y n^P \tag{4.35}$$

where $Y = j + 1$. Similar to the above arguments, to assess the damage we need to have some knowledge of the critical damage at a vibration stress. Let's assume this occurs at N_1 at stress level G_1. Then, as in Equation (4.29), the thermodynamic damage ratio at any other stress G_2 level at n_2 cycle is

$$\text{Vibration damage} = \left(\frac{n_2}{N_1}\right)^p \left(\frac{G_2}{G_1}\right)^Y. \tag{4.36}$$

If damage is 1, $n_2 = N_2$, and failure occurs. Following the arguments of the other examples, we find the acceleration factor is

$$\text{AF}_{\text{damage}} = \left(\frac{N_1}{N_2}\right) = \left(\frac{G_2}{G_1}\right)^b \tag{4.37}$$

where $b = Y/P$. Since the number of cycles is related to cycle frequency f and the time τ according to

$$N = f\tau, \tag{4.38}$$

then if f is constant $\text{AF}_{\text{damage}}$ is a commonly used relationship for cyclic compression where

$$\text{AF}_{\text{vibration}} = \frac{N_1}{N_2} = \frac{\tau_1}{\tau_2} = \left(\frac{G_1}{G_2}\right)^{-b}_{\text{sinusoidal}} \equiv \left(\frac{G_{\text{rms}1}}{G_{\text{rms}2}}\right)^{-b}_{\text{random}}. \tag{4.39}$$

This is commonly used for the acceleration factor in sinusoidal testing. For random vibration above, we substitute for G the random vibration G_{rms} level [13] (see Special Topics B for examples).

It is helpful to write the linear form for cycles to failure for a particular stress. This is deduced to within a constant from the above equation

$$N_1 = A(G_1)^{-b} \equiv A(G_{\text{rms}1})^{-b} \tag{4.40}$$

where $A = N_2/G_2^{-b}$ is treated as a constant. This is essentially the relation that holds for what is called the *S–N* curves (see Figure 1.3). Note that if we write the cyclic equation with $G \propto S$ where S is the stress, we have

$$N_1 = CS_1^{-b} \text{ or } S_1 = KN_1^{-B} \tag{4.41}$$

where $B = 1/b$, C is the proportionality constant when going from G to S, and K is a constant similar to C. Note that some authors write this as proportional to the strain instead of the stress. The relationship is generally used to analyze *S–N* data. b and C are often referred to as Basquin's equation, commonly written as

$$N = CS^{-b}. \tag{4.42}$$

Note that since stress and strain are conjugate variables, one can write this alternately in terms of the strain. However, it is usually written this way as the experimental number of cycles to failure N for a given stress S level constitutes *S–N* curve data in fatigue testing of materials. Such data are widely available in the literature. The slope of the *S–N* curve (see Figure 1.3) provides an estimate of the exponent b above. *S–N* data are commonly determined using sinusoidal stress. Often we do not know b. Typical values of b are $4 < b < 8$. Some guidance is provided in MIL-STD-810F. For example, it recommends $b = 8$ for broadband random (discussed below) and $b = 6$ where for profiles. A conservative test is obtained by setting $b = 4$. Testing can of course can determine b for a particular failure mode.

Most device vibration testing is typically either sinusoidal or random. The goal is to try and accelerate the type of vibration occurring under use conditions. For automotive, for example, this is random vibration. For a piece of equipment undergoing cyclic motion, this is more likely sinusoidal. The relation for random data is $b/2$ when using the power spectral density (PSD) (G_{rms}^2/Hz) level instead of G_{rms}. This is evident from the fact that the G_{rms} level is found as the square root of the area under the PSD spectrum:

$$G_{rms} = \sqrt{\int_{f_1}^{f_2} \text{PSD}(f)df}. \tag{4.43}$$

In the simple case of random vibration, white noise for example, we have

$$G_{rms} \cong \sqrt{W_{PSD}\Delta f}. \tag{4.44}$$

Here the time compression expression above for random vibration is over the same bandwidth $\Delta f_1 = \Delta f_2$, then inserting Equation (4.44) into Equation (4.39) we have

$$\text{AF}_W = \frac{\tau_1}{\tau_2} = \left(\frac{W_{PSD1}}{W_{PSD2}}\right)^{-b/2} = \frac{N_1}{N_2}. \tag{4.45}$$

This is a common form used for the random vibration acceleration factor [13] (see Special Topics B for examples). The general form for the cycles to failure for a particular stress is then

$$N_{\text{failure}} \equiv \beta (W_{\text{PSD}})^{-b/2}. \tag{4.46}$$

Finally, it should be noted that if we have applied a number of different stresses, the general damage ratio can be written by accumulating the thermodynamic damage along the same work path (Equation (4.36)) as

$$\text{Damage} = \sum_i \left(\frac{n_i}{N_i}\right)^p = \sum_i \left(\frac{n_i}{\text{AF}_{\text{vibration}}(1,i)N_1}\right)^p. \tag{4.47}$$

This equation turns out to be extremely useful. In Section 4.4 and Special Topics B.7 we illustrate how this equation can be used for environmental profiling in order to find a CAST goal.

4.3.5 Example 4.6: Cycles to Failure under a Resonance Condition: Q Effect

When a product has a built-in resonance that is causing a higher stress level, the fatigue life will be shortened due to magnification of the resonance. In order to model the resonance condition for fatigue life, we first need to explain some common resonance terms to aid the unfamiliar reader. Resonance is commonly quantified using an amplitude magnification factor denoted Q which is typically measured in a number of ways. Figure 4.4 illustrates a resonance. The most common way to measure a resonance is through the transmissibility. The transmissibility is the output divided by the input peak G or G_{rms} level. This is illustrated for sinusoidal vibration in Figure 4.4, giving

$$Q \equiv \frac{\text{Output amplitude}(G_s)}{\text{Input amplitude}(G_s)} = \frac{5.2}{1} = 5.2. \tag{4.48}$$

An alternate way to measure Q when the resonance curve is available but the input level is not known is by assessing the resonance value f_0 divided by the resonance width Δf at

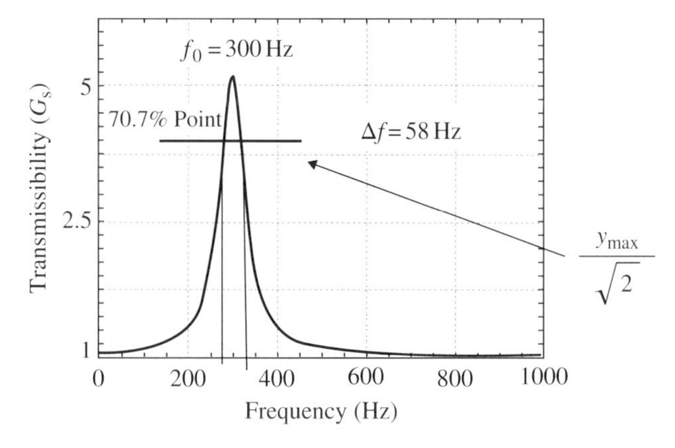

Figure 4.4 Graphical example of a sine test resonance

the maximum amplitude, divided by the square root of 2 as illustrated in Figure 4.4, and given by

$$Q \equiv \frac{f_0}{\Delta f} = \frac{300\ \text{Hz}}{52\ \text{Hz}} = 5.2. \tag{4.49}$$

Since we have already quantified the cycles to failure in terms of the G or G_{rms} level, then under a resonance condition the output G level is simply multiplied by the Q magnification level at resonance so the time to failure is

$$N_{\text{failure}} = \frac{1}{\beta_0}\left(QG_{\text{input}}\right)^{-b} \equiv \frac{1}{\beta_0}\left(G_{\text{rms}}(Q)\right)^{-b}. \tag{4.50}$$

Here $G_{\text{rms}}(Q)$ indicates that the G_{rms} level is a function of Q. We see the cycles to failure generally goes as the input G-level multiplied by Q for sinusoidal vibration to an inverse power, so that larger Q and/or G values yield a smaller number of cycles to failure. Recall that since time is related to cycles through the frequency of vibration as time = N-cycles/frequency, then we can put this in terms of time to failure if needed.

If we have a random vibration input then the G_{rms} level is not easily assessed. The simplest instructional way to estimate the G_{rms} level from the input PSD level at resonance is to use the well-known Miles' equation [14, 15], defined:

$$G_{\text{rms acc}}(Q) = \sqrt{\frac{\pi}{2} f_0 Q W_{\text{PSD-input}}}. \tag{4.51}$$

Miles' equation is applicable for a single degree of freedom. It is an approximate formula that assumes a flat power spectral density in the neighborhood of the resonance f_0. As a rule of thumb, it may be used if the power spectral density is flat over at least two octaves centered at the natural frequency. To ensure narrow band resonance over a flat portion of the spectrum, a $Q \sim 10$ or higher is often recommended.

As an example for using the Miles' equation, if we have a single degree of vibration system with an 80 Hz resonance having a Q of 12 and a random 0.04 $G_{\text{rms}}{}^2$/Hz input level then the G_{rms} resonance level at resonance is estimated as:

$$G_{\text{rms}} = \sqrt{\frac{\pi}{2} 80(12)0.04} = 7.8\, G_{\text{rms}}. \tag{4.52}$$

Random vibration is not simple to understand. As the term suggests, the input vibration levels are not constant but are random in nature. Therefore, there is a probability of occurrence in amplitude. The Miles' equation above is written in terms of a 1-sigma response which means then when the random vibration is roughly Gaussian that this Grms level occurs about 68.27% of the time. Since Miles' equation stipulates that the PSD be approximately flat near the resonance condition, then this is atypical for a Gaussian-like input in the frequency domain.

Using Miles' equation, the time to failure can therefore be written for a single degree of freedom where the natural frequency response to a random vibration input can be approximated as

Table 4.2 Damping loss factor examples for certain materials

Material	Damping loss factor (η)
Metals	<0.01
Steel	0.001–0.002
Aluminum	0.007–0.005
Rubber (depends on type)	0.01–0.05
Frictional spring	0.1–0.5

above for a $Q \sim 10$ or higher and a flat PSD input near f_0 and, under these conditions, will occur about 67% of the time (1-sigma) as

$$N_{\text{failure}} = \frac{1}{\beta_0}\left(\frac{\pi}{2} f_0 Q W_{\text{PSD input}}\right)^{-b/2}. \tag{4.53}$$

It is important to note that Q is the inverse of damping as

$$Q = 1/(2\xi) = 1/\eta \tag{4.54}$$

where ξ is the damping factor and η is the loss factor. Often materials are characterized using the loss factor [3] and it is helpful to know which materials can be used to reduce resonances. Table 4.2 provides a list of typical well-known loss factors for different materials.

4.4 Cumulative Damage Accelerated Stress Test Goal: Environmental Profiling and Cumulative Accelerated Stress Test (CAST) Equations

A very important use of the cumulative stress equations in this chapter is in accelerated testing goals by doing proper environmental profiling of fielded stress conditions. These equations can help as they already define the cumulative stress damage equations. To that end, we would like to define a new term here called the CAST goal. The CAST goal depends on the work path, that is, the type of stress, and the total fielded use time τ_1 (or N_1 cycles) needed to design a test for 10 years or 20 000 cycles, for example. We may not know the time to failure N_1 or t_1, but we may only need 10 years of life or equivalent cycles. If we survive this equivalent damage time on an accelerated test, then we have proven that our product can survive the fielded use time it is designed for by doing a properly designed accelerated test using the CAST goal.

The issue is that a product in the field is typically exposed to varying stresses over time, so how do we specify one CAST goal and at what equivalent stress level? We do this by cumulating all the i stresses and associated times at each ith environmental stress relative to one particular reference stress of interest we call Stress 1. If this were say temperature with numerous estimated temperature profiles in the field, we collapse it to something like 10 years at 50°C. Obviously if there is only one constant temperature, for example 50°C, that a product is exposed to in the field, we do not need to cumulate the fielded environments and collapse them as we already know the fielded stress and simply define the time goal of say 10 years. Table 4.3 provides some very common CAST goals for different fielded stress work path types. Note that

Table 4.3 Cumulative stress test goals: CAST equations

Stress (work path)	Original equation	CAST equations and goal* (survival time or no. cycles)
Fatigue (Miner's Rule)	(4.11)	$N_1 = \sum_i \left(\dfrac{n_i}{AF_{damage}(1,i)} \right)$
Creep	(4.19)	$\tau_1 = \sum_i \left(\dfrac{t_i}{AF_{creep}(1,i)} \right)^p$
Wear	(4.26)	$\tau_1 = \sum_i \left(\dfrac{t_i}{AF_{wear}(1,i)} \right)$
Thermal cycle	(4.33)	$N_1 = \sum_i \left(\dfrac{n_i}{AF_{cyclic\ fatigue}(1,i)} \right)^p$
Vibration	(4.33)	$N_1 = \sum_i \left(\dfrac{n_i}{AF_{vibration}(1,i)} \right)^p$
Temperature (corrosion)	(5.10)	$\tau_1 = \sum_i \left(\dfrac{t_i}{AF_{temperature}(1,i)} \right)$

*AF_{damage} is the damage acceleration factor defining the work path.

we have included Equation (5.10) from the next chapter for completeness. To illustrate how to use CAST goals, an example is provided in Section B7.1.

4.5 Fatigue Damage Spectrum Analysis for Vibration Accelerated Testing

In vibration theory, the method for assessing damage has broadened to the use of spectral information. Often we would like to accelerate field data in vibration testing to simulate product life in the lab in a short time frame. However, field data spectrums are difficult to reproduce and accelerate correctly as, in reality, a product experiences lots of different types of vibration during its life. A great tool was developed to help reproduce potential field data damage to representative field spectra, called fatigue damage spectrum (FDS). Once FDS field data are obtained, a test engineer can statistically simulate the end-use environment to test their product. Once the end-use environment is simulated, the test engineer can accelerate a test to a desired test duration value more accurately. In addition, a test engineer can predict the life expectancy of a product by adjusting the target life to produce a desired spectral excitation for a test. Such FDS software is now available [16].

FDS is based on Miner's Rule for damage where fatigue damage accumulates in a product until it fails. In FDS theory, using Miner's rule the total damage a product experiences in a particular time can be calculated from field data and plotted for a specific range of frequencies. The resulting plot of fatigue damage versus frequency is the FDS, a means by which to quantify the stress–strain loads placed on a product.

Field data are collected by accelerometers that record the vibration levels at a number of positions on the product in the field under use condition of concern (i.e., likely worst-case fatigue conditions). The fatigue damage dosage for each maneuver is calculated using an

FDS, which effectively plots damage versus frequency. The damage from each field use condition of concern is summed over the usage profile of the product to estimate the likely life accumulated damage. From this profile we determine a statistically representative vibration test which contains at least the same damage content as the product's lifetime, but over a short test period.

FDS theory exists for both sine and random vibration testing. We start with sine vibration as it is typically easier to illustrate, then we explain it for random vibration.

The use of random vibration for testing is often used instead of sine vibration as it more often than not represents real-world situations. Random vibration is mostly analyzed by a spectrum in the frequency domain rather than the time domain. Random vibration is then as the name indicates: random. Its actual distribution depends on test requirements.

4.5.1 *Fatigue Damage Spectrum for Sine Vibration Accelerated Testing*

In sine vibration, we create cyclic fatigue. The input G level will cause a stress on the components and can cause cyclic fatigue that we described in Equation (3.23). The G stress level even in sine vibration will vary with frequency, and Equation (3.25) cyclic damage can be written

$$\text{Cyclic damage}(f) = \frac{\sum_{n} \oint Y_n(f) dX_n}{W_{\text{failure}}} = \frac{\sum_{f_{\min}}^{f_{\max}} \oint G_{n,f} dX_n}{W_{\text{failure}}}. \tag{4.55}$$

If we knew the cyclic work area at each test frequency point, perhaps with the use of a strain gage, we could accumulate the amount of work that was done in some period of time where cycles $n = f \times$ (test time). Often it is approximated using Miner's rule, so we can write the damage over a resonance point due to frequency test from $f_{\min} < f_n < f_{\max}$ as:

$$\text{Cyclic damage}(f) = \text{FDS}(f_n) = \sum_{f\min}^{f_{\max}} d_i(f) = \sum_{f_{\min}}^{f_{\max}} \frac{n_i}{N_{f,i}} \tag{4.56}$$

where d_i is the damage at the ith stress level which varies with frequency. This of course is an alternate form of Miner's Rule (Equation (4.9)). For example,

$$d_i = \frac{n_i}{N_i} = n_i \frac{S^b}{C} \tag{4.57}$$

where we have used Equation (4.42) ($N = CS^{-b}$) as in Equation (4.11). The stress level will go as the vibration amplitude, denoted here as $Z(f)$ (function of frequency f) to within a K factor, $S(f) = KZ(f)$, so we write the fatigue damage spectra as [17–20]

$$\text{FDS}(f_n) = \frac{K^b}{C} \sum_i n_i Z_i^b(f). \tag{4.58}$$

4.5.1.1 Example 4.7: Fatigue Damage Spectrum for Sine Vibration at and across Resonance

At resonance, we only have one stress level i so that n simply equates to $n = f_0 t$ (where t is the test time). Next in sine testing the displacement is defined $X = Z\sin(\omega t)$, where ω_0 is the resonance frequency. The acceleration is the second derivative so the maximum acceleration = $Z\omega_0{}^2$ and the acceleration in terms of G_s is gG, so that $gG = Z\omega_0{}^2$. However, the peak to peak displacement is what is needed so this is a factor of 2 difference; furthermore, since we are at resonance we must take into account the amplification factor Q at resonance (see Figure 4.4) so $Z = 2gGQ/\omega_0{}^2$. Then Miner's rule for fatigue testing at a sinusoidal resonance yields the following FDS spectrum:

$$\mathrm{FDS}(f_0) = n\frac{K^b}{C}\left(\frac{2gQG}{\omega_0^2}\right)^b = f_0 t\frac{K^b}{C}\left(\frac{2gQG}{\omega_0^2}\right)^b. \tag{4.59}$$

If we are testing across the resonance, then the amplitude is given by

$$\mathrm{FDS}(f_0) = f_0 t\frac{K^b}{C}\frac{(gG)^b}{\omega_0^{2b}\left[\left(\omega_0^2 - \omega^2\right)^2 - \left(\dfrac{\omega\omega_0}{Q}\right)^2\right]^{b/2}}. \tag{4.60}$$

The denominator is found from the harmonic oscillator amplitude equation which is found in numerous books and articles. It is not reproduced here as it is easily referenced.

4.5.2 *Fatigue Damage Spectrum for Random Vibration Accelerated Testing*

In a very similar manner we can use the concepts for random vibration. We will consider looking at the general damage spectrum (GDS) near a resonance (that will be related to FDS shortly) in random vibration. Then we will have, similar to Equation (4.58),

$$\mathrm{GDS}(f_n) = \frac{K^b}{C} f_0 t Z_i^b(f). \tag{4.61}$$

In random vibration FDS theory, the stress is proportional to the relative displacement of the single degree of freedom (SDOF) (i.e., single axis) multiplied by a constant K:

$$Z \approx \sigma_f = K\, G_{\mathrm{rms\,disp}}(f_0). \tag{4.62}$$

When the test vibration distribution is Gaussian, Mile's equation is in many cases a good estimate for the G_{rms} displacement that takes place. That is, there is a certain probability for displacement amplitudes. Using Mile's equation (Equation (4.51)), this is:

$$G_{\mathrm{rms\,disp}}(f_0) = \frac{gG_{\mathrm{rms\,acc}}(f_0)}{(2\pi f_0)^2} = \frac{9.8}{(2\pi f_0)^2}\left(\frac{\pi}{2} f_0 Q W_{\mathrm{PSD\text{-}input}}(f)\right)^{1/2}. \tag{4.63}$$

By inserting Equations (4.63) and (4.62) into (4.61), we have for the general damage fatigue spectrum for any ith SDOF narrow-band random vibration Gaussian

$$\mathrm{GDS}(f_i) = \frac{K^b}{C} f_i t \left(\frac{9.8}{(2\pi f_i)^2} \right)^b \left(\frac{\pi}{2} f_i Q W_{\text{PSD-input}}(f) \right)^{b/2}. \tag{4.64}$$

Finally, the FDS differs from the GDS theory by what is called the expected values for the amplitude for a narrow-band Gaussian, which only differs by a gamma numeric factor that goes with b as shown below

$$\mathrm{FDS}(f_i) = \frac{K^b}{C} f_i t \left(\frac{9.8}{(2\pi f_i)^2} \right)^b \left(\frac{\pi}{2} f_i Q W_{\text{PSD-input}}(f) \right)^{b/2} \Gamma\left(1 + \frac{b}{2}\right). \tag{4.65}$$

These data in the field are collected over the ith $W_{\text{PSD-input}}$ narrow-band Gaussian areas of frequency, as these are worst cases. Then the FDS is generated for the environment. In general, the constants are taken as $K = C = 1$, the exponent $b = 4$, 8, or 12 (see discussion in Section 4.3.4) and the amplification factor $Q = 10$, 25, or 50. The inverse of this equation reproduces the test spectrum and is written

$$W(f_i) = \frac{2(2\pi f_i)^3}{9.8^2 Q} \left(\frac{C\, FDS(f_i)}{K^b f_i t \Gamma\left(1 + \frac{b}{2}\right)} \right)^{2/b}, \; i = 1, 2, \ldots, N. \tag{4.66}$$

Once we have the test profile, $W(f)$ the actual test-level $W(f)$-Test is specified by changing the t to t_{eq} for the test to accelerated time. This is consistent with Equation (4.46) when time is put in terms of cycles to failure. Typical values for b or G according to the Mil-Std 810F are around 8.

Summary

Thermodynamic work is perhaps the most directly measurable and practical quantity to use in assessing a system's free energy and its degradation.

4.2 Example 4.1: Miner's Rule

Miner's rule assumes Work$(\omega, n) = n$ Work(σ) for cyclic fatigue damage. Using this approximation, we obtain an effective damage

$$\text{Effective damage} \approx \frac{\sum_{i=1}^{K} n_i w_i}{W_{\text{failure}}}. \tag{4.4}$$

For stress strain for example,

$$\begin{aligned}\text{Effective damage} &\approx \frac{n_1\oint\sigma_1 de}{W_{\text{failure}}}+\frac{n_2\oint\sigma_2 de}{W_{\text{failure}}}+\frac{n_3\oint\sigma_3 de}{W_{\text{failure}}}+\ldots \\ &= \frac{n_1\oint\sigma_1 de}{N_1 W_1}+\frac{n_2\oint\sigma_2 de}{N_2 W_2}+\frac{n_3\oint\sigma_3 de}{N_3 W_3}+\ldots\end{aligned} \tag{4.7}$$

Therefore, Miner's rule for cumulative fatigue damage becomes

$$\text{Effective damage} \approx \frac{n_1}{N_1}+\frac{n_2}{N_2}+\frac{n_3}{N_3}+\ldots=\sum_{i=1}^{k}\frac{n_i}{N_i} \tag{4.9}$$

and Miner's rule approximation for cumulative fatigue damage with an acceleration factor is

$$\text{Effective damage} = \sum_{i=1}^{k}\frac{n_i}{\text{AF}_{\text{damage}}(1,i)N_1}. \tag{4.11}$$

4.3.1 *Example 4.2: Creep Cumulative Damage and Acceleration Factors*

Using the work methodology, we found creep damage ratio as

$$\text{Creep damage} = \frac{w}{W_{\text{failure}}} = e^{-Q/k_{\text{B}}\left(\frac{1}{T_2}-\frac{1}{T_1}\right)}\left(\frac{\sigma_2}{\sigma_1}\right)^{M}\left(\frac{t_2}{t_1}\right)^{p}. \tag{4.15}$$

The creep acceleration factor was obtained as

$$\text{AF}_{\text{creep}} = \left(\frac{\tau_2}{\tau_1}\right) = e^{-E_{\text{a}}/k_{\text{B}}\left(\frac{1}{T_1}\ \frac{1}{T_2}\right)}\left(\frac{\sigma_1}{\sigma_2}\right)^{K}. \tag{4.16}$$

The creep time to failure was found as

$$\ln(t_{\text{failure}}) = C + \frac{E_{\text{a}}}{K_{\text{B}}T} - K\ln(\sigma). \tag{4.18}$$

For a number of different i stresses along the same work path,

$$\text{Damage} = \sum_i\left(\frac{t_i}{\tau_i}\right)^{p} = \sum_i\left(\frac{t_i}{\text{AF}_{\text{creep}}(1,i)\,\tau_1}\right)^{p}. \tag{4.19}$$

The equation turns out to be extremely useful and its use is detailed in Special Topics B.7 and B.7.1. There, we provide examples of how to use this equation for environmental profiling in accelerated test planning.

4.3.2 Example 4.3: Wear Cumulative Damage and Acceleration Factors

Using the work methodology, we found wear damage ratio as

$$\text{Wear damage} = \frac{w_2}{W_1} = \frac{\left(k_P \frac{P_{W2}\nu_2 t}{HA_2}\right)}{\left(k_P \frac{P_{W1}\nu_1 \tau_1}{HA_1}\right)} = \frac{\left(P_{W2}\frac{\nu_2 t}{A_2}\right)}{\left(P_{W1}\frac{\nu_1 \tau_1}{A_1}\right)}. \tag{4.22}$$

The wear acceleration factor was

$$\mathrm{AF_{wear}}(2,1) = \frac{\tau_1}{\tau_2} = \left[\frac{\left(P_{W2}\frac{\nu_2}{A_2}\right)}{\left(P_{W1}\frac{\nu_1}{A_1}\right)}\right]_{\mathrm{DF}}. \tag{4.23}$$

Time to failure for wear was obtained as

$$t_{\text{failure}} = C\frac{(A)}{(P_{W1}\nu_1)}. \tag{4.25}$$

Accumulating damage at each stress level can also be written for wear as

$$\text{Damage} = \sum_i \left(\frac{t_i}{\tau_i}\right) = \sum_i \left(\frac{t_i}{\mathrm{AF_{wear}}(1,i)\tau_1}\right). \tag{4.26}$$

The equation turns out to be extremely useful and its use is detailed in Special Topics B.7 and B.7.1. There, we provide examples of how to use this equation for environmental profiling in accelerated test planning.

4.3.3 Example 4.4: Thermal Cycle Fatigue and Acceleration Factors

Using the work methodology, we found thermal cyclic fatigue damage ratio as

$$\text{Thermal cycle damage} = e^{-E_a/k_B\left(\frac{1}{T_2}-\frac{1}{T_1}\right)}\left(\frac{\Delta T_2}{\Delta T_1}\right)^M \left(\frac{n_2}{N_1}\right)^p. \tag{4.29}$$

The cyclic fatigue acceleration factor was obtained

$$\mathrm{AF}_{\text{cyclic fatigue}} = \left(\frac{N_2}{N_1}\right) = e^{-E_a/k_B\left(\frac{1}{T_1}-\frac{1}{T_2}\right)}\left(\frac{\Delta T_1}{\Delta T_2}\right)^K. \tag{4.30}$$

Thermal cycles to failure equation was found as

$$\ln(N_{\text{failure}}) = C + E_a/Tk_B + K\ln(\Delta T). \tag{4.32}$$

Accumulating thermodynamic damage per cycle

$$\text{Damage} = \sum_i \left(\frac{n_i}{N_i}\right)^p = \sum_i \left(\frac{n_i}{\mathrm{AF}_{\text{cyclic fatigue}}(1,i)N_1}\right)^p. \tag{4.33}$$

The equation turns out to be extremely useful and its use is detailed in Special Topics B.7 and B.7.1 where we provide examples of how to use this equation for environmental profiling in accelerated test planning.

4.3.4 Example 4.5: Mechanical Cycle Vibration Fatigue and Acceleration Factors

Using the work damage methodology, the sine vibration damage ratio was found as

$$\text{Vibration damagc} = \left(\frac{n_2}{N_1}\right)^p \left(\frac{G_2}{G_1}\right)^Y. \tag{4.36}$$

Acceleration factor as a function of G level or G_{rms} level for both sinusoidal or random vibration could be written as

$$\mathrm{AF}_{\text{vibration}} = \frac{N_1}{N_2} = \frac{\tau_1}{\tau_2} = \left(\frac{G_1}{G_2}\right)^{-b}_{\text{sinusoidal}} \equiv \left(\frac{G\ \text{rms}_1}{G_{\text{rms}2}}\right)^{-b}_{\text{random}}. \tag{4.39}$$

Mechanical vibration cycles to failure model was obtained as

$$N_{\text{failure}} = \frac{1}{\beta_0}(G)^{-b} \equiv \frac{1}{\beta_0}(G_{\text{rms}})^{-b}. \tag{4.40}$$

The time compression for random vibration was written in terms of W_{PSD} spectra

$$\mathrm{AF}_W = \frac{\tau_1}{\tau_2} = \left(\frac{W_{\text{PSD1}}}{W_{\text{PSD2}}}\right)^{-b/2} = \frac{N_1}{N_2}. \tag{4.45}$$

The general form for the cycles to failure for a particular stress then shows

$$N_{\text{failure}} \equiv \beta (W_{\text{PSD}})^{-b/2}, \tag{4.46}$$

accumulating the thermodynamic damage along the same work path as

$$\text{Damage} = \sum_i \left(\frac{n_i}{N_i}\right)^p = \sum_i \left(\frac{n_i}{\text{AF}_{\text{vibration}}(1,i)N_1}\right)^p. \tag{4.47}$$

The equation turns out to be extremely useful and its use is detailed in Special Topics B.7 and B.7.1. There, we provide examples of how to use this equation for environmental profiling in accelerated test planning.

4.3.5 *Example 4.6: Cycles to Failure under a Resonance Condition: Q Effect*

For a Q magnification level at resonance, the cycles to failure is

$$N_{\text{failure}} = \frac{1}{\beta_0}(QG_{\text{input}})^{-b} \equiv \frac{1}{\beta_0}(G_{\text{rms}}(Q))^{-b} \tag{4.50}$$

where Q is the inverse of damping, defined

$$Q = 1/(2\xi) = 1/\eta, \tag{4.54}$$

and $G_{\text{rms}}(Q)$ indicates that the G_{rms} level is a function of Q, defined using the Miles' equation as

$$G_{\text{rms}}(Q) = \sqrt{\frac{\pi}{2} f_0 Q W_{\text{PSD-input}}}. \tag{4.51}$$

For a random vibration spectrum, the cycles to failure can be written

$$N_{\text{failure}} = \frac{1}{\beta_o}\left(\frac{\pi}{2} f_0 Q W_{\text{PSD input}}\right)^{-b/2}. \tag{4.53}$$

4.4 Cumulative Damage Accelerated Stress Test Goal: Environmental Profiling and Cumulative Accelerated Stress Test (CAST) Equations

Equations (4.11), (4.19), (4.26), and (4.33) as well as (5.10) in the following chapter can help define CAST goal.

4.5 Fatigue Damage Spectrum Analysis for Vibration Accelerated Testing

FDS theory uses the Miner's rule to estimate the total damage a product experiences over a particular time, calculated from field data and plotted for a specific range of frequencies. The resulting plot of fatigue damage versus frequency is the FDS.

4.5.1 Fatigue Damage Spectrum for Sine Vibration Accelerated Testing

If we knew the cyclic damage approximated using Miner's rule over a sine resonance point in a frequency test from $f_{\min} < f_n < f_{\max}$, then

$$\text{Cyclic damage}(f) = \text{FDS}(f_n) = \sum_{f\min}^{f_{\max}} d_i(f) = \sum_{f_{\min}}^{f_{\max}} \frac{n_i}{N_{f,i}}. \tag{4.56}$$

Then we found that Miner's rule for fatigue testing at a sinusoidal resonance yields the following FDS spectrum

$$\text{FDS}(f_0) = n\frac{K^b}{C}\left(\frac{2gQG}{\omega_0^2}\right)^b = f_0 t\frac{K^b}{C}\left(\frac{2gQG}{\omega_0^2}\right)^b. \tag{4.59}$$

If we are testing across the resonance, then the amplitude is given by

$$\text{FDS}(f_0) = f_0 t\frac{K^b}{C}\frac{(gG)^b}{\omega_0^{2b}\left[\left(\omega_0^2 - \omega^2\right)^2 - \left(\frac{\omega\omega_0}{Q}\right)^2\right]^{b/2}}. \tag{4.60}$$

4.5.2 Fatigue Damage Spectrum for Random Vibration Accelerated Testing

The FDS for any ith SDOF narrow-band random vibration Gaussian is given by

$$\text{FDS}(f_i) = \frac{K^b}{C}f_i t\left(\frac{9.8}{(2\pi f_i)^2}\right)^b \left(\frac{\pi}{2}f_i Q W_{\text{PSD-input}}(f)\right)^{b/2}\Gamma\left(1+\frac{b}{2}\right). \tag{4.65}$$

The inverse of this equation reproduces the test spectrum and is given by

$$W(f_i) = \frac{2(2\pi f_i)^3}{9.8^2 Q}\left(\frac{C\,FDS(f_i)}{K^b f_i t\Gamma\left(1+\frac{b}{2}\right)}\right)^{2/b}, \ i = 1, 2, \ \ldots., N. \tag{4.66}$$

References

[1] Feinberg, A. and Widom, A. (2000) On thermodynamic reliability engineering. *IEEE Transaction on Reliability*, **49** (2), 136.

[2] Feinberg, A. Using Thermodynamic Work for Determining Degradation and Acceleration Factors. ASTR 2015 (available also at http://www.dfrsoft.com, accessed 5 May 2016).

[3] Feinberg, A. (2015) Thermodynamic damage within physics of degradation, in *The Physics of Degradation in Engineered Materials and Devices* (ed J. Swingler), Momentum Press, New York.

[4] Miner, M.A. (1945) Cumulative damage in fatigue. *Journal of Applied Mechanics*, **12**, A159–A164.

[5] Collins, J.A., Busby, H. and Staab, G. (2010) *Mechanical Design of Machine Elements and Machines*, 2nd edn, John Wiley & Sons, Inc., New York.

[6] Archard, J.F. (1953) Contact and rubbing of flat surface. *Journal of Applied Physics*, **24** (8), 981–988.

[7] Archard, J.F. and Hirst, W. (1956) The wear of metals under unlubricated conditions. *Proceedings of the Royal Society*, **A-236**, 397–410.

[8] Hirst, W. (1957) *Proceedings of the Conference on Lubrication and Wear*, Institution of Mechanical Engineers, London, p. 674.

[9] Coffin, L.F. (1954) A study of the effects of cyclic thermal stresses on a ductile metal. *Transactions of the ASME*, **76**, 923–950.

[10] Coffin, L.F. (1974) Fatigue at high temperature: prediction and interpretation. James Clayton Memorial Lecture. *Proceedings of the Institution of Mechanical Engineers (London)*, **188**, 109–127.

[11] Manson, S.S. (1953) *Behavior of Materials under Conditions of Thermal Stress*, NACA-TN-2933 from NASA, Lewis Research Center, Cleveland.

[12] Norris, K.C. and Landzberg, A.H. (1969) Reliability of controlled collapse interconnections. *IBM Journal of Research and Development*, **13** (3), 266–71.

[13] MIL-STD-810G. 2008. Method 514.6, Military Standard 810G, Annex A, 31 October 2008.

[14] Miles, J.W. (1954) On structural fatigue under random loading. *Journal of the Aeronautical Sciences*, **21**, 753.

[15] Steinberg, D.S. (2000) *Vibration Analysis for Electronic Equipment*, Wiley-Interscience, New York.

[16] Vibration Research Corporation is one good source that provides FDS software and helps implement the process. See http://www.vibrationresearch.com, accessed on 4 May 2016.

[17] Downing, S.D. and Socie, D.F. (1982) Simple rainflow counting algorithms. *International Journal of Fatigue*, **4**, 31–40.

[18] Bishop, N.W.M. and Sherratt, F. (1989) Fatigue life prediction from power spectral density data. Part 2: recent development. *Environmental Engineering*, **2** (1 and 2), 5–10.

[19] Halfpenny, A., Kihm, F. Mission profile and testing synthesis based fatigue damage spectrum. Proceedings of the 9th International Fatigue Congress, 2006, Atlanta.

[20] McNeill, S.I. Implementing the Fatigue Damage Spectrum and Fatigue Equivalent, Vibration Testing, Sound and Vibration. Proceedings of the 79th Shock and Vibration Symposium, October 26–30, 2008, Orlando, FL, pp. 1–20.

5

Corrosion Applications in NE Thermodynamic Degradation

5.1 Corrosion Damage in Electrochemistry

We have noted that the Helmholtz free energy equated to work at constant temperature (T) and sometimes we also used constant volume (V) (see Equation (2.99)). Similarly, for constant temperature and pressure (P) corrosion processes, the change in the Gibbs free energy G bounds the ability to do useful work (see Section 2.11.3).

$$W \le G_{\text{final}} - G_{\text{initial}}.$$

In detail, the change in the Gibbs free energy is given by

$$\Delta G = \Delta U + P\Delta V - T\Delta S.$$

From the first law we have that the internal energy is the heat minus any type of work, which we describe as compressive and non-compressive:

$$\Delta U = Q - P\Delta V - W_{\text{non-compressive}}.$$

Then substituting this equation into ΔG for quasistatic isothermal processes $Q \le T\Delta S$ we can write as in Equation (2.108)

$$\Delta G \ge W_{\text{non-compressive}}. \tag{5.1}$$

Thermodynamic Degradation Science: Physics of Failure, Accelerated Testing, Fatigue, and Reliability Applications, First Edition. Alec Feinberg.

Therefore, the Gibbs free energy bounds the work process. In actuality, for many corrosion processes we will be able to determine the corrosion damage from the thermodynamic work done as they are reasonably quasistatic, so that

$$\Delta W = -\Delta G + T\Delta S_{\text{damage}}.$$

The equality for a reversible process in Equation (5.1) indicates that the free energy is the reversible work. As we found in Chapter 3, we can therefore interpret the above equation as a statement for the thermodynamic work in the form

$$W_{\text{actual}} = W_{\text{rev}} - W_{\text{irr}}.$$

In this section we provide some examples of how to use the concept of work damage in electrochemistry. We include examples for Miner's rule for secondary batteries, chemical corrosion processes, corrosion current in primary batteries, and corrosion rate in microelectronics. In Chapter 1 we described four main types of aging. Corrosion is in the category of a complex aging process. It often involves thermal activation, forced electrical excitation, and can also include mechanical issues such as stress corrosion cracking. We can often separate these by looking at the rate-controlling mechanism. There are a number of ways to look at corrosion processes in electrochemistry. We start with an unusual treatment in the following example by applying a Miner's rule application.

5.1.1 Example 5.1: Miner's Rule for Secondary Batteries

Miner's rule is not limited to mechanical stress; it can be applied to many types of cyclic fatigue situations.

For example, one interesting application is to chemical cells of secondary batteries [1, 2]. Here the cyclic work from Table 1.1 for k charge–discharge cycles is

$$w(\text{cycle}) = \sum_{i=1}^{k} \oint_{\text{Area}_i} V dQ. \tag{5.2}$$

In an analogous manner, battery manufacturers plot something similar to S–N curves found in mechanical stress–strain application. This is depth of discharge percent (DoD%) versus charging–discharging cycles to failure N. However, by comparison, in this case the DoD strain variable, charge is plotted using the DoD% (which is a percent of the charging capacity). This is plotted instead of the stress cycles to failure by battery manufacturers for an apparent stress failure threshold of voltage as shown in Figure 5.1.

In an application of Miner's rule for secondary batteries using such available data the sum is over the DoD% ith level for battery life pertaining to a certain failure (permanent) voltage drop (such as 10%) of the initially rated battery voltage. Then the effective damage done in n_i DoD%

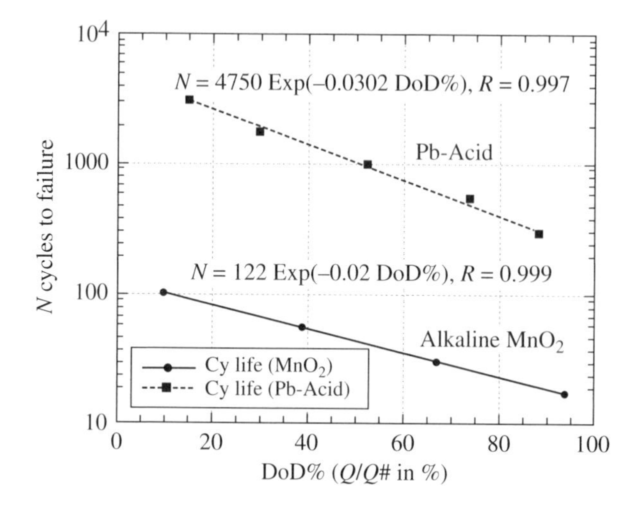

Figure 5.1 Lead acid and alkaline MnO_2 batteries fitted data. Source: Feinberg and Widom [2]

can be assessed when N_i is known for the ith DoD level. For k-types of DoD% for various ith levels, Miner's rule for secondary battery takes on the familiar form

$$\text{Battery damage} \approx \sum_{i=1}^{k} \frac{n_i(\text{DoD})}{N_i(\text{DoD})} \tag{5.3}$$

where n and N are related to the number of DoD taken for the ith DoD level. The S–N type of DoD% curve is modeled [1, 2] as chemical rate where activation plays a role in battery N cycle life:

$$N_j = N_o e^{-\frac{\Phi_j}{K_B T}} = N_{jo} e^{-f_j \,\text{DoD\%}}. \tag{5.4}$$

This is plotted in Figure 5.1. Here we assume the activation free energy is a linear function of depth of discharge [1, 2] (see Equation (5.26) and Chapter 6 for details of the activation free energy):

$$\Phi_j = \phi_j + \nu Q_j \tag{5.5}$$

where ν is the activation damage voltage; Q_j is the value for which the battery manufacturer assumes the chemical cell to be fully discharged for the jth battery type (see Equation (5.2)); and ϕ is a damage activation energy. We then define

$$N_{jo} = N_0 e^{-\frac{\phi_j}{k_B T}}, \quad \text{and} \quad f_j \text{DoD\%} = \frac{\nu Q_j}{k_B T}. \tag{5.6}$$

5.2 Example 5.2: Chemical Corrosion Processes

In a corrosion process we have from Table 1.1 that the Gibbs work is $\mu\, dN$ where we have N_i particles (or mole) of species i reacting. If the total reaction is consumed after N_i particles, and the reaction proceeds to the point where only $M_i < N_i$ particles have reacted, then we can immediately determine the damage from the thermodynamic work. The resulting work done on the chemically reacting system may be described in terms of the *chemical potential* μ_i of the species k. That is, under constant temperature and pressure, the work is directly related to the Gibbs free energy change so the damage is defined, from Table 1.1 and Equation (2.114),

$$\text{Damage} = \frac{w_{\text{done}}}{W_{\text{failure}}} = \frac{\Delta G_{M \text{ particles}}}{\Delta G_{N \text{ particles}}} = \frac{\sum_i \mu_i M_i}{\sum_i \mu_i N_i}. \tag{5.7}$$

However, chemical potentials are not easily assessed. Further, this says very little about the rate of the reaction. The expression is therefore not easily quantified. We need to work with measurable parameters that are available. For example, we may know the corrosion rate or corrosion current under a certain stress condition. Then instead of using the chemical potential approach to assess damage, we can default to the thermodynamic work rule. For example, using the corrosion current (Table 1.1), the work is $\Delta W = VIdt$. A primary battery is an example where the V and I are known and the battery degradation is due to chemical cell corrosion. If V–I is not a rate-controlling stress and temperature is instead, we will show how to use this. Although we do not have cyclic work, we can think of it in terms of corrosion time for a single cycle. Then considering the corrosion time, we write the corrosion damage

$$\text{Damage (corrosion)} = \sum_{i=1}^{n} \frac{(VI)_i t_i}{(VI)_i \tau_i} = \frac{(VI)_1 t_1}{(VI)_1 \tau_1} + \frac{(VI)_2 t_2}{(VI)_2 \tau_2} + \ldots = \sum_{i=1}^{n} \frac{t_i}{\tau_i} \tag{5.8}$$

where for the ith stress condition (e.g., temperature stress i), t_i is the corrosion time and τ_i time to fail. We find the expression simplifies and we do not actually need to know the corrosion current–voltage at any stress level if we know the time to failure τ_i at each ith stress level, similar to Miner's rule. We can view this for general, galvanic, or specific corrosion processes such as primary batteries.

It can be difficult to always know the failure time at any ith stress level. However, since we are on the same work path, we can often establish an acceleration factor between stress levels. If AF_C (C for corrosion acceleration factor) and r_1 are known, then the above equation can be simplified for the ith stress using

$$\tau_i = AF_{\text{corr damage}}(1, i)\tau_1.$$

The term $AF_{\text{corr damage}}$ will be related to the corrosion rate which, in the most general form from above, will be between the ith and jth stress environment. Similar to how we found the acceleration factors in Chapter 4, using Equation (5.7) we find

$$\mathrm{AF}_{\text{corr damage}}(i,j) = \frac{(VI(T))_i}{(VI(T))_j}. \tag{5.9}$$

Note that the corrosion current in general will also be a function of the ith stress temperature T_i. That is, temperature may need to be added to the acceleration factor (see Equation (5.20)). Below we develop a number of other expressions for $\mathrm{AF}_{\text{damage}}$ that may be of interest. Then the damage equation for corrosion can be simplified as

$$\text{Damage (corrosion)} = \sum_{i=1}^{k} \frac{t_i}{\mathrm{AF}_{\text{corr damage}}(1,i)\tau_1}. \tag{5.10}$$

The equation turns out to be extremely useful. In Section 4.4 and Special Topics B.7 we illustrate how this equation can be used for environmental profiling in order to find a cumulative accelerated stress test (CAST) goal in the form of

$$\text{Cum } \tau_1 \text{ profile} = \sum_{i=1}^{k} \frac{t_i}{\mathrm{AF}_{\text{corr damage}}(1,i)}. \tag{5.11}$$

When our product is subjected to multiple stresses in the field, we will find a certain amount of corrosion damage at τ_1 (say 5 years). If we wanted to design an accelerated test to simulate this damage, how do we select one equivalent stress? Equation (5.11) will allow us to profile and find one overall representative equivalent stress that would equate to this same amount of damage observed at τ_1. Once we have one equivalent stress for the damage amount at τ_1 we have an environmental profile and can design an accelerated test that will accelerate time and create the same damage amount in a shorter period of test time as observed in the field (at 5 years). An example is provided in Special Topics B.7.1.

Now consider the case when the damage value is 1; failure occurs and we can derive a useful expression for the acceleration factor in terms of the stress. However, in this case we will first need an expression for the corrosion rate.

A common expression for the corrosion rate in terms of mass transferred in the reaction [3] is given according to Faraday's Law (dM/dt) as proportional to the net current $I(T)$, a function of the reaction rate in temperature (discussed below) as

$$dM = k\,I(T)\,dt \tag{5.12}$$

where dM is the mass of metal dissolving (g) and

$$k = \left(\frac{A_m}{nH}\right) = \frac{\text{atomic wt of metal}}{\text{no. electrons transferred} \times 96\,500\text{ A s}}. \tag{5.13}$$

Thus, A_m is the atomic mass. The corrosion rate (dM/dt) is then

$$\text{Corrosion rate} = \left(\frac{A_m}{nH}\right) I(T). \tag{5.14}$$

5.2.1 *Example 5.3: Numerical Example of Linear Corrosion*

It can be confusing to use the equation. As a simple example, consider iron corroding in air-free acid at an electrochemical corrosion rate of 1 μA/cm^2. It dissolves as ferrous ions (Fe^{+2}) and therefore $n = 2$. We can obtain the corrosion rate in mils per year (mpy) using the above expression

$$\text{Corrosion rate} = \frac{55.8\ \text{g/equivalent}\left(10^{-6}\ \text{C/s-cm}^2\right)}{2(96\,487\ \text{C/equivalent})} = 2.89 \times 10^{-10}\text{g/s-cm}^2. \tag{5.15}$$

To convert the corrosion rate to mpy, first divide by the density of iron (7.86 g/cm^3). Additionally there are 3.15×10^7 s/year and 393.7 mils/cm (=1 inch/2.54 cm × 1000 mils/inch), then

$$\text{Corrosion rate} = 2.89 \times 10^{-10}\text{g/s-cm}^2 \times \frac{1}{7.86\ \text{g/cm}^3} \times 3.15 \times 10^7\text{s/year} \times 393.7\ \text{mils/cm} \tag{5.16}$$

$$\text{Corrosion rate} = 0.456\ \text{mpy}.$$

Here mpy is a common corrosion rate unit. Table 5.1 provides relative values for estimating corrosion resistance of materials. From the table, note that 0.46 mpy is an outstanding corrosion rate, indicating an excellent corrosive resistive material. Note this table of relative values is established under a certain set of environmental conditions.

When the corrosion rate is linear, the rate can be found by simply dividing the mass corroded by the corrosion time. In units of R (cm/h), this is the mass W with exposed area A in a laboratory experiment, defined

$$R(\text{cm/h}) = \frac{W(\text{g})}{t(\text{h})D(\text{g/cm}^3)A(\text{cm}^2)}. \tag{5.17}$$

Here, the expression can be checked using its units. Now converting to a mixed unit expression, the conversion factor R(cm/h)(3.44×10^6 mils/cm × h/year) gives R in mpy. Thus, R(cm/year) = 1/(3.44×10^6) R(mpy). Since A(cm^2)(inch^2/(2.54 cm)^2) gives A in inch^2, then A(cm^2) = 6.45 A(inch^2). Similarly, W(g)/1000 = W(mg). Inserting these values into the above equation, we have

Table 5.1 Estimated relative corrosion resistance

Relative corrosion resistance	Mils per year (mpy)	Micrometer per year (μm/year)
Outstanding	<1	<25
Excellent	1–5	25–100
Good	5–20	100–500
Fair	20–50	500–1000
Poor	50–200	1000–5000
Unacceptable	200+	5000+

$$R(\text{mpy}) = \frac{3.44\times10^6 \dfrac{W(\text{mg})}{1000}}{t(\text{h})D(\text{g/cm}^3)6.45A\left(\text{inch}^2\right)} = \frac{534\,W(\text{mg})}{t(\text{h})D(\text{g/cm}^3)A\left(\text{inch}^2\right)}. \tag{5.18}$$

This is one common mixed unit expression used in corrosion engineering.

5.2.2 *Example 5.4: Corrosion Rate Comparison of Different Metals*

Using Faraday's law we can make a comparison of the corrosion amount and rate for different metals. Consider an experimental set-up as shown in Figure 5.2. Let the anode be various metals. The cathode can be say stainless steel (iron). Let us consider a corrosion current flowing of 5 mA for 1 year. Table 5.2 provides the calculated results using Faraday's linear law of corrosion.

As an example of one of the calculations, the amount of metal removed for iron in 1 year at 5 mA is 45.6 g. This is found as

$$\begin{aligned}45.6\,\text{g} &= [55.8\ \text{g/eq.}/(96\ 500\,\text{C/equivalent}\times 2)]\times(5\,\text{mA}/1000)\\ &\quad\times(365\,\text{Days}\times 24\,\text{h}\times 60\times 60).\end{aligned}$$

If we were trying to find an electrode with the least amount of corrosion mass, we note that titanium is the best choice. However, if cost was a consideration, stainless steel (iron) would likely be the second-best choice.

5.2.3 *Thermal Arrhenius Activation and Peukert's Law*

The corrosion current is dependent on the actual situation, but in general is thermally activated across a free energy barrier

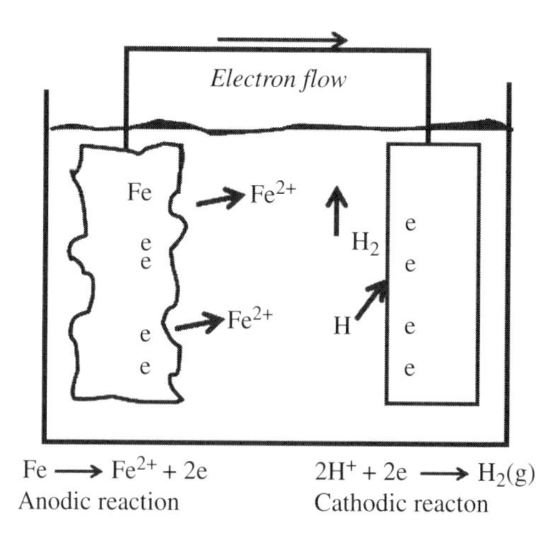

Figure 5.2 A simple corrosion cell with iron corrosion

Table 5.2 Predicted corrosion rates and amounts for 1 year at 5 mA of current for anodic different metals [7]

Metal	Atomic weight (g/mol)	Outer shell electrons	Density (g/cc)	Rate (mpy)	Corrosion amount (g)
Copper	63.55	1	8.96	4558	103.8
Iron	55.8	2	7.874	2277	45.6
Brass*	64.15	2	8.3561	2467	52.4
Titanium	47.88	2	4.54	3389	39.1
Nickel	58.69	2	8.9	2119	48.0
Tin	118.71	4	7.31	2609	48.5
Zinc	65.39	2	7.13	2947	53.4
Indium	114.8	3	7.31	3364	62.5
Lead	207.2	4	11.35	2933	84.7
Tungsten	183.85	2	19.25	3069	150.2
Iridium	192.22	2	22.4	2757	157.1
Silver	107.86	1	10.5	6602	176.3

*67Cu.33Zn.

$$I(T) = I_0 \exp\left\{-\frac{E_R}{RT}\right\} = I_0 \exp\left\{-\frac{E_a}{k_B T}\right\}. \tag{5.19}$$

This is the well-known Arrhenius dependence which can be written either using R, the gas constant, given as 8.31 J/K mole and E_R is in joules, or using k_B, the Boltzmann constant (8.6173×10^{-5} eV/K) with E_a in electron-volts (eV). The gas constant is more commonly used in electrochemistry than the Boltzmann constant. I_0 is current rate constant for generalized corrosion current. We can now consider two different environments at different temperatures. If we have obtained a damage value of 1 between these two stress environments, then from Equations (5.8) and (5.19):

$$\text{Damage (corrosion)} = 1 = \frac{(VI)_1 \tau_1}{(VI)_2 \tau_2}, \text{ and}$$
$$\text{AF} = \frac{\tau_2}{\tau_1} = \frac{\text{Rate}(T_1)}{\text{Rate}(T_2)} = \frac{I_1}{I_2} = \exp\left\{-\frac{\Delta\varepsilon}{R}\left(\frac{1}{T_1} - \frac{1}{T_2}\right)\right\} = \exp\left\{-\frac{E_a}{k_B}\left(\frac{1}{T_1} - \frac{1}{T_2}\right)\right\}. \tag{5.20}$$

Here we have let $V_1 = V_2$ since temperature often dominates the corrosion process. This is the well-known Arrhenius acceleration factor (see Special Topics B for application of this factor). The rate for any one stress is

$$\text{Rate}(T) = \nu \exp\left\{-\frac{E_a}{k_B T}\right\} \tag{5.21}$$

where ν is a rate constant (which could be written in terms of another rate using the acceleration factor in Equation (5.20) if known, e.g., $\text{Rate}(T_1) = \text{Rate}(T_2)\,\text{AF}$). We note that E_a is often thought of as a barrier height. The barrier height is a property of the system's free energy related to the type of material (see Chapter 6 for more details). The expression is often written in linear $Y = ax + b$ form so that the time to failure can be plotted for regression analysis as

$$\ln(t_{\text{failure}}) = C + E_a/k_B T \tag{5.22}$$

where C is simply $\ln(1/\nu)$.

Note that the voltage potential usually does not factor in different stress temperatures; only the current (often a function of voltage) will likely change or dominate in the corrosion process so

$$\text{Damage (corrosion)} = \frac{I_1 t_1}{I_2 \tau_2} = \text{AF}(1,2)\frac{t_1}{\tau_2}. \tag{5.23}$$

The reader should note that in general the current rate constant I_0 cancels out but, depending on the way Equation (5.19) is written and the test conditions, this may not always be the case.

An alternate expression for primary batteries is based on Peukert's law [3–5]

$$C_p = I^Y t \tag{5.24}$$

where C_p is the battery capacity at a one-ampere discharge rate (expressed in Ampere-hours); I is the discharge current in Amperes; Y is the Peukert constant (typically between 1.05 and 1.15); and t is the time of discharge in hours. The acceleration factor for two different cell types would then be

$$\text{AF} = \frac{\tau_2}{\tau_1} = \frac{C_{p2}}{C_{p1}}\left(\frac{I_1}{I_2}\right)^Y. \tag{5.25}$$

5.3 Corrosion Current in Primary Batteries

Even simple corrosion is a complex aging process. A chemical battery is similar to that shown in Figure 5.2, which exemplifies some fundamentals of the process. This illustrates the four elements necessary for corrosion to occur: a metal anode; cathode; electrolyte; and a conductive path. If any one of these is removed, corrosion can be prevented.

Figure 5.3 illustrates how one can visualize this same electrochemical general corrosion process on a simple metal surface when an electrolyte is present. Small irregularities on the surface can form cathode C and anode A areas, usually due to differences in oxygen concentration.

The exchange of matter can be described in non-equilibrium thermodynamics in terms of the currents at the electrodes. Corrosion involves two separate processes of half-reactions, oxidation, and reduction. At the anode, oxidation reaction consumes metal atoms when they corrode, releasing electrons. These electrons are used up in the reduction reaction at the cathode. A common expression for the corrosion current in aqueous corrosion is from the Butler–Volmer equation [4] which gives the anode and cathode currents I_a, I_c respectively for each electrode

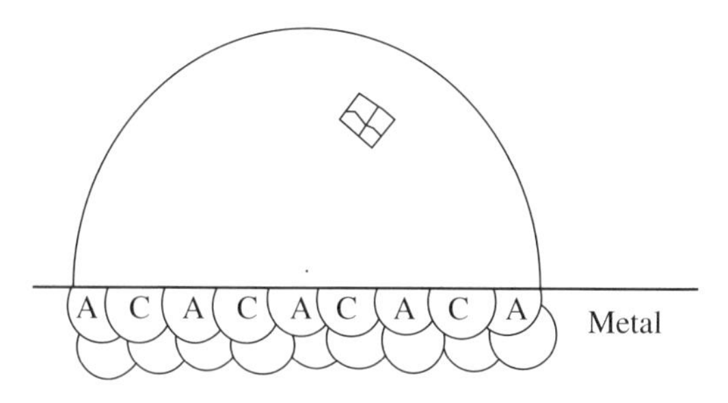

Figure 5.3 Uniform electrochemical corrosion depicted on the surface of a metal

$$i_{corr} = i_a - i_c = I_{0a} \exp\left(\frac{\Delta G_a}{RT}\right) - I_{0c} \exp\left(\frac{\Delta G_c}{RT}\right). \tag{5.26}$$

Here we have identified the barrier height in terms of the Gibbs free energy. The thermodynamic work to change the free energy or equivalently the barrier height equates to $\Delta G = -W = -qE$ where $q = zH$; H is the Faraday constant; and z is the stoichiometric number of electrons in the reaction, that is,

$$M \Leftrightarrow M^{Z+} + ne \tag{5.27}$$

where M is the metal-forming M^{Z+} ions in solution. The anode and cathode work is distributed (i.e., $n_a = n\alpha$, $n_c = n(1-\alpha)$, often $\alpha = 0.5$) so that the anode work amount is for example

$$\Delta G_a = -\alpha n H E. \tag{5.28}$$

For $E > 0$ the reaction is spontaneous and no applied potential is required for the reaction to occur. In the case of a battery in steady state with reasonable current flow, anodic reaction can dominate. The AF would be modified, so if a non-spontaneous reaction is forced the potential is expressed in the acceleration factor as

$$\text{AF} \approx \frac{I_{f2}}{I_{f1}} = \exp\left[-\frac{n\alpha HE}{R}\left(\frac{1}{T_2} - \frac{1}{T_1}\right)\right]. \tag{5.29}$$

5.3.1 Equilibrium Thermodynamic Condition: Nernst Equation

When $I_a = I_c = I_0$ there is no net corrosion current and the equilibrium condition yields

$$C_{0a} K_a^0 \exp\left(-\frac{n\alpha HE}{RT}\right) = C_{0c} K_c^0 \exp\left(\frac{(1-\alpha)nHE}{RT}\right). \tag{5.30}$$

Here we substitute into Equation (5.26) the anode and cathode amplitudes [3] $I_{0a} = C_{0a}K^0{}_a\, nFA$ and $I_{0c} = C_{0c}K^0{}_c\, nFA$. $K^0{}_a$ and $K^0{}_c$ are temperature-dependent rate constants, $C_0 = C_{0a}$ is the concentration of the reducing agent at the anode and $C_R = C_{0c}$ is that of the oxidizing agent at the cathode electrode surface.

Collecting terms yields the famous Nernst thermodynamic equilibrium condition

$$E = E^0 + \frac{RT}{nH}\ln\left(\frac{C_R}{C_0}\right) \quad \text{or} \quad \Delta G = \Delta G^0 - RT\ln\left(\frac{C_R}{C_0}\right) \tag{5.31}$$

where the ratio C_R/C_0 is often called the reaction quotient. E^0 is the standard open-circuit cell potential and ΔG^0 is the standard free energy, defined

$$E^0 = \frac{RT}{nH}\ln\left(\frac{K_f^0}{K_b^0}\right), \quad \Delta G^0 = nHE^0 = RT\ln\left(\frac{K_f^0}{K_b^0}\right). \tag{5.32}$$

The Nernst equation enables the calculation of the thermodynamic electrode potential when concentrations are known. It can also indicate the corrosive tendency of the reaction. When the thermodynamic free energy of the process is negative, there is a spontaneous tendency to corrode (as we have discussed in Section 2.11.3).

5.4 Corrosion Rate in Microelectronics

In aqueous corrosion (above) the currents are easier to measure than they are in microelectronics, occurring arbitrarily on a circuit board. In this instance, the concept of adding the local percent relative humidity (%RH) at or near the surface has been found to aid in describing the potential for corrosion to occur. The rate of corrosion and the rate of mass transport are related then to the local %RH present at the surface which enhances the electrolyte at the surface for conducting the corrosion currents. The corrosion current is not well defined, but is proportional to the rate kinetics and this local %RH as

$$I_{\text{corr}} \propto (\%\text{RH})^M K(T). \tag{5.33}$$

In microelectronics, failures due to corrosion are accelerated under higher-temperature and -humidity conditions than normally occur during use. In accelerated testing, the acceleration factor between the testing stress and use environment, having different temperature and humidity conditions, can be found from the ratio of the currents:

$$\text{AF}_{TH} = \frac{I_{\text{stress}}}{I_{\text{use}}} = \left(\frac{\text{RH}_{\text{stress}}}{\text{RH}_{\text{use}}}\right)^M \left(\frac{K(T)_{\text{stress}}}{K(T)_{\text{use}}}\right) = \text{AF}_H\,\text{AF}_T. \tag{5.34}$$

This is the temperature–humidity acceleration factor used in humidity testing, with examples provided in Special Topics B. The temperature acceleration factor was found in Equation (5.20) and is defined for each factor as:

$$\text{AF}_T = \exp\left\{\frac{E_a}{k_B}\left[\frac{1}{T_{\text{use}}} - \frac{1}{T_{\text{stress}}}\right]\right\}, \quad \text{and} \quad \text{AF}_H = \left(\frac{\text{RH}_{\text{stress}}}{\text{RH}_{\text{use}}}\right)^M \tag{5.35}$$

where E_a is the activation energy related to the failure mechanism. This is called the Peck's acceleration model [6]. Typically, E_a and M are not found in testing, but are estimated based on historical data; often values of $M = 2.66$ and $E_a = 0.7$ eV are used. We note that in microelectronic failure due to corrosion this accelerated factor can be a large number as it has two components $\mathrm{AF}_T \times \mathrm{AF}_H$ (see examples of its use in Special Topics B).

From Equation (5.33), we can write the corrosion current with stress as

$$\frac{1}{I_{\text{stress}}} = C\left(\mathrm{RH}_{\text{use}}\right)^{-M}\left(K(T)_{\text{use}}\right)^{-1}. \tag{5.36}$$

The time to failure goes as $t_{\text{failure}} \sim 1/I_{\text{stress}}$ and, combining the above with Equations (5.22) and (5.34), we have for the time to failure

$$\ln\left(t_{\text{failure}}\right) = C_0 + \frac{E_a}{k_B T} - M\ln\left(\mathrm{RH}_{\text{stress}}\right). \tag{5.37}$$

5.4.1 Corrosion and Chemical Rate Processes Due to Temperature

Here we provide more details of the simple overview in Section 5.4. In microelectronics, surface corrosion is a function of the local %RH. The rate of corrosion and the rate of mass transport is related to the local %RH present at the surface. The reaction rate depends on concentration C according to the *chemical differential rate law* (also the law of mass action [4, 5]). For example, in reaction

$$3A + 2B \rightarrow D + E \tag{5.38}$$

the differential rate law may have concentrations $[A]$, $[B]$, and $[D]$. The rate is defined

$$K(C) = \frac{d[D]}{dt} = k[A]^n[B]^m \tag{5.39}$$

where n and m are power exponents with orders of A and B, respectively. The square brackets, such as those around $[A]$, indicate the concentration in moles per liter of A; $K(C)$ is then the reaction rate as a function of concentration. The overall order of the reaction (n and m) cannot be predicted from the reaction equation but must be found experimentally.

In terms of microelectronics, surfaces have an affinity for local %RH near the surface. This feeds the thin-film electrolyte which affects the reaction rate, both in terms of concentration in the anodic and cathodic reactions and also in terms of the rate of mass transport. For many corroding metals the cathodic reduction of water itself in the electrolyte ($2H_2O \rightarrow OH^- + H_2 - e^-$) is a rate-controlling process [2, 3].

The overall chemical reaction rate, if it could be described in simple terms as provided above, is thus some function of the local %RH. We will therefore use a somewhat naive approach by assuming that the %RH, similar to the formulation for $K(C)$ in the chemical differential rate law, has an overall rate that goes as a power functional form with the local %RH itself as

$$K\{C\} \propto (\mathrm{rh})^m \tag{5.40}$$

where rh = RH/100. We will see that this assumption is consistent with corrosion kinetics of Peck's expression where Peck's expression is applicable [6]. Applicability may be as high as greater than 60%RH, depending upon the corrosion occurring. Some metals, such as iron, do not corrode below a certain %RH value. To show that this expression is consistent with Peck's expression, we first insert this assumption for $K(C)$ into our expressions for the corrosion currents as

$$I_{\text{forward}} = (\text{rh})^m nHAK_{\text{forward}}C_0 \quad \text{and} \quad I_{\text{backward}} = (\text{rh})^m nHAK_{\text{backward}}C_R \tag{5.41}$$

(note that when rh = 1, the original expression results). The net current $I = I_{\text{forward}} + I_{\text{backward}}$ is zero under equilibrium conditions. For situations where the net current is not zero, the net current approaches that of either the forward or backward current, depending on the dominating mechanism. For example, anodic corrosion usually dominates a corrosion process. In this case, I is approximately I_{forward} and the corrosion current is (Table 5.2)

$$I = (\text{rh})^m \, n\, HAC_0K_{\text{forward}} \exp\left\{ -\frac{\Delta G_{\text{forward}}}{RT} \right\}. \tag{5.42}$$

In accelerated testing the acceleration factor between a stress and use environment, having different temperature and humidity conditions, can be found from the rate ratio

$$\text{AF}_{\text{TH}} = \frac{\text{Corrosion rate}_{\text{stress}}}{\text{Corrosion rate}_{\text{use}}} = \frac{\left(\frac{A_m}{nH}\right) I_{\text{stress}}}{\left(\frac{A_m}{nH}\right) I_{\text{use}}} = \frac{I_{\text{stress}}}{I_{\text{use}}}. \tag{5.43}$$

This is another way of writing Equation (5.34), where $\Delta \text{G}_{\text{forward}}/R = E_a/k_B$ and

$$\text{AF}_{TH} = \text{AF}_T \text{AF}_H \tag{5.44}$$

as described in Equation (5.35).

Summary

5.1 Corrosion Damage in Electrochemistry

The Gibbs free energy bounds the work process. For many corrosion processes, we will be able to determine the corrosion damage from the thermodynamic work done as they are reasonably quasistatic, so that

$$\Delta W = -\Delta G + T\Delta S_{\text{damage}}.$$

The equation is a statement for the thermodynamic work in the form

$$W_{\text{actual}} = W_{\text{rev}} - W_{\text{irr}}.$$

5.1.1 Example 5.1: Miner's Rule for Secondary Batteries

In an analogous manner, battery manufacturers plot something similar to *S–N* curves found in mechanical stress–strain application. This is DoD% versus charging–discharging cycles to failure *N*. Using Table 1.1, the analogy for strain variable charge (proportional to the stress variable voltage) is replaced with DOD%, and *N* for secondary batteries is then the number of charge–discharge cycles to failure. The threshold for failure is the degraded voltage that the battery can be charged to, as shown in Figure 5.1.

$$\text{Battery damage} \approx \sum_{i=1}^{k} \frac{n_i(\text{DoD})}{N_i(\text{DoD})} \tag{5.3}$$

where n and N are related to the number of DoD taken for the ith DoD level.

5.2 Example 5.2: Chemical Corrosion Processes

In a corrosion process we may have N_i particles (or mole) of species i reacting. If the total reaction is consumed after N_i particles, and the reaction proceeds to the point where only $M_i < N_i$ particles have reacted, then we can immediately determine the damage from the thermodynamic work.

$$\text{Damage} = \frac{\Delta G_{M \text{ particles}}}{\Delta G_{N \text{ particles}}} = \frac{\sum_i \mu_i M_i}{\sum_i \mu_i N_i}. \tag{5.7}$$

For example, using the corrosion current the work is $\delta W = VIdt$. Although we do not have cyclic work, we can think of it in terms of corrosion time for a single cycle. Then, considering the corrosion time, we write the corrosion damage

$$\text{Damage (corrosion)} = \sum_{i=1}^{n} \frac{(VI)_i t_i}{(VI)_i \tau_i} = \frac{(VI)_1 t_1}{(VI)_1 \tau_1} + \frac{(VI)_2 t_2}{(VI)_2 \tau_2} + \ldots = \sum_{i=1}^{n} \frac{t_i}{\tau_i}. \tag{5.8}$$

Then the damage equation for corrosion can be simplified as

$$\text{Damage (corrosion)} = \sum_{i=1}^{k} \frac{t_i}{\text{AF}_{\text{corr damage}}(1,i)\tau_1}. \tag{5.10}$$

A common expression for the corrosion rate in terms of mass transferred in the reaction [3] is given according to Faraday's Law (dM/dt) as proportional to the net current $I(T)$

$$\text{Corrosion rate} = \left(\frac{A_m}{nH}\right) I(T). \tag{5.14}$$

5.2.3 Thermal Arrhenius Activation and Peukert's Law

The corrosion current is dependent on the actual situation, but in general is thermally activated across a free energy barrier

$$I(T) = I_0 \exp\left\{-\frac{E_R}{RT}\right\} = I_0 \exp\left\{-\frac{E_a}{k_B T}\right\}. \tag{5.19}$$

An alternate expression for primary batteries is based on Peukert's law [3–5]:

$$C_p = I^Y t. \tag{5.24}$$

The acceleration factor for two different cell types would then be

$$\text{AF} = \frac{\tau_2}{\tau_1} = \frac{C_{p2}}{C_{p1}}\left(\frac{I_1}{I_2}\right)^Y. \tag{5.25}$$

5.4 Corrosion Rate in Microelectronics

The corrosion current is not well defined but is proportional to the rate kinetics and this local %RH as

$$I_{\text{corr}} \propto (\%\text{RH})^M K(T). \tag{5.33}$$

In accelerated testing the acceleration factor between the testing stress and use environment, having different temperature and humidity conditions, can be found from the ratio of the current

$$\text{AF}_{TH} = \frac{I_{\text{stress}}}{I_{\text{use}}} = \left(\frac{\text{RH}_{\text{stress}}}{\text{RH}_{\text{use}}}\right)^M \left(\frac{K(T)_{\text{stress}}}{K(T)_{\text{use}}}\right) = \text{AF}_H \, \text{AF}_T. \tag{5.34}$$

This is the temperature–humidity acceleration factor used in humidity testing with examples provided in Special Topics B. We have for each factor:

$$\text{AF}_T = \exp\left\{\frac{E_a}{k_B}\left[\frac{1}{T_{\text{use}}} - \frac{1}{T_{\text{stress}}}\right]\right\}, \quad \text{and} \quad \text{AF}_H = \left(\frac{\text{RH}_{\text{stress}}}{\text{RH}_{\text{use}}}\right)^M \tag{5.35}$$

The time to failure goes as $t_{\text{failure}} \sim 1/I_{\text{stress}}$ and, combining the above with Equation (5.22), we have

$$\ln(t_{\text{failure}}) = C_0 + \frac{E_a}{k_B T} - M \ln(\text{RH}_{\text{stress}}). \tag{5.37}$$

References

[1] Feinberg, A. and Widom, A. (2000) On thermodynamic reliability engineering. *IEEE Transaction on Reliability*, **49** (2), 136.
[2] Feinberg, A., Widom, A. Thermodynamic extensions of Miner's Rule to chemical cells. In *Proceedings of Annual Reliability and Maintainability Symposium*, January 24–27, 2000, Los Angeles, CA, pp. 341–344.
[3] Linden, D. (ed) (1980) *Handbook of Batteries and Fuel Cells*, McGraw-Hill, New York.
[4] Fontana, M.G. and Greene, N.D. (1978) *Corrosion Engineering*, McGraw-Hill, New York.
[5] Uhlig, H.H. (1967) *Corrosion and Corrosion Control*, McGraw-Hill, New York.
[6] Peck, D.S. Comprehensive model for humidity testing correlation. In *Proceedings of the 24th IEEE International Reliability Physics Symposium*, April 1986, Anaheim, CA, pp. 44–50.
[7] DfRSoftware, http://www.dfrsoft.com, author's software, see A.1.1 for details.

6

Thermal Activation Free Energy Approach

6.1 Free Energy Roller Coaster

Thermally activated processes are one of the four main types of aging described in Chapter 1. We now ask: how does the free energy change as degradation work occurs? Sometimes, the system path to the free energy minimum is smooth and downhill all the way to the bottom. For other systems, the path may descend to a relative minimum, but not an absolute minimum, something resembling a roller coaster. The path goes downhill to what looks like the bottom and faces a small uphill region. If that small hill could be scaled, then the final drop to the true minimum would be just over the top of the small hill. The small climb before the final descent to the true minimum is called a free energy barrier. The system may stay for a long period of time in the relative minimum before the final decay to true equilibrium.

Often the time spent in the neighborhood of the relative minimum is the lifetime of a fabricated product, and the final descent to the true free energy minimum represents the catastrophic failure of the product.

The estimated lifetime τ over which the system stays at the relative minimum obeys the Arrhenius law in Equation 5.21 as $(1/\tau) = 1/\tau_0 \exp(-\Delta\phi/k_B T)$ where $\Delta\phi$ is the height of the free energy barrier (ϕ_c).

The activation energy can be thought of as the amount of energy needed for the degradation process. This defines a special relative equilibrium aging state

$$\phi_B \leq \phi \leq \phi_B + \Delta\phi \tag{6.1}$$

where the subscript B refers to the barrier and $\Delta\phi$ is the barrier height.

Thermodynamic Degradation Science: Physics of Failure, Accelerated Testing, Fatigue, and Reliability Applications, First Edition. Alec Feinberg.

6.2 Thermally Activated Time-Dependent (TAT) Degradation Model

When activation is the rate-controlling process, Arrhenius-type rate kinetics applies. In this section, the parametric time-dependence of an Arrhenius mechanism is addressed. Parametric degradation is often termed "soft failure" or "device drift over time," for example slow power loss of a transistor or creep change. This thermal activation mechanism explains logarithmic-in-time aging of many key device parameters described here. Such a mechanism, with temperature as the fundamental thermodynamic stress factor, leads to this predictable logarithmic-in-time-dependent aging on measurable parameters. Here the thermodynamic work is evident in the free energy path.

There are two thermally activated time-dependent (TAT) device degradation models [1, 2]. If degradation is graceful over time, then many instances occur in which logarithmic-in-time aging results for one or more device parameters. This aging and examples are described below. Degradation leading to catastrophic failure using the TAT model is presented at the end of this chapter, while applications for the TAT model are provided in the next chapter.

In Arrhenius processes, the probability rate *dp*/*dt* to surmount the relative minimum free energy barrier ϕ as in Figure 6.1 can be written

$$\frac{dp}{dt} = \nu \exp\left(-\frac{\phi}{k_B T}\right) \tag{6.2}$$

where ν is a rate constant; k_B is Boltzmann's constant; T is the temperature; and t is time. More information on this probability distribution is given in Section 6.3. We wish to associate the thermodynamic aging kinetics in the material with measurable parametric changes. Therefore, we model the above as a fractional rate of parametric change given by

$$\frac{da}{dt} = \nu \exp\left(-\frac{\phi(a)}{k_B T}\right) \tag{6.3}$$

where a is the fractional change of the measurable parameter. We can let a be dimensionless, defined $a = \Delta P / P_0$. For example, ΔP could be a parameter change that is of concern such as resistance change, current change, mechanical creep strain change, voltage transistor gain change, and so forth, such that a is then the fractional change.

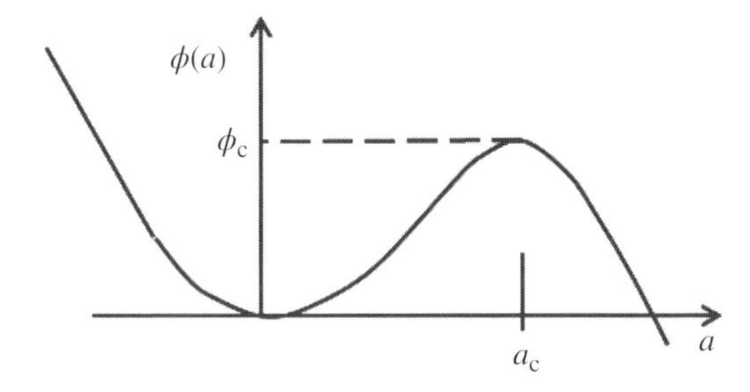

Figure 6.1 Arrhenius activation free energy path having a relative minimum as a function of generalized parameter a. Source: Feinberg and Widom [1]. Reproduced with permissions of IEEE

In the TAT model the aging process is closely related to the parametric change. The assumption is that the free energy itself will be associated with the parameter through the thermodynamic work.

Thus, ϕ will be a function of a as indicated in Equation (6.3). Then the free energy can be expanded in terms of its parametric dependence using a Maclaurin series (with environmental factor held constant for the moment). The free energy is defined:

$$\phi(a) = \phi(0) + ay_1 + \frac{a^2}{2}y_2 + \dots \tag{6.4}$$

where y_1 and y_2 are given by

$$y_1 = \frac{\partial\phi(0)}{\partial a} \quad \text{and} \quad y_2 = \frac{\partial^2\phi(0)}{\partial a^2}. \tag{6.5}$$

6.2.1 Arrhenius Aging Due to Small Parametric Change

When $a << 1$, the first and second terms in the Maclaurin series yield

$$\frac{da}{dt} = \nu(T)\exp\left(-\frac{ay_1}{k_B T}\right) \tag{6.6}$$

where

$$\nu(T) = \nu_0 \exp\left(-\frac{\phi(0)}{k_B T}\right).$$

Rearranging terms, and solving for a as a function of t and integrating, provides a *logarithmic-in-time aging TAT model* where

$$a = \frac{\Delta P}{P} \cong A\ln(1 + Bt) \quad \text{for } a << 1, \tag{6.7}$$

where A and B are defined

$$A = \frac{k_B T}{y_1} \quad \text{and} \quad B = \frac{\nu(T)y_1}{k_B T}. \tag{6.8}$$

Logarithmic-in-time aging is an important process since the origin of this aging kinetics can mathematically be tied to the Arrhenius mechanisms of which numerous examples exist [1, 2]. Figure 6.2 illustrates typical logarithmic-in-time aging. One notes that aging is highly non-linear for early time. This curve is representative of many aging and kinetic processes such as crystal frequency aging drift [3], corrosion of thin films, chemisorption processes [4], early degradation of primary battery life [5], activated creep (see Section 7.3), activated wear (see Section 7.2), transistor key parameter aging (see Section 7.4), and so forth. The significance

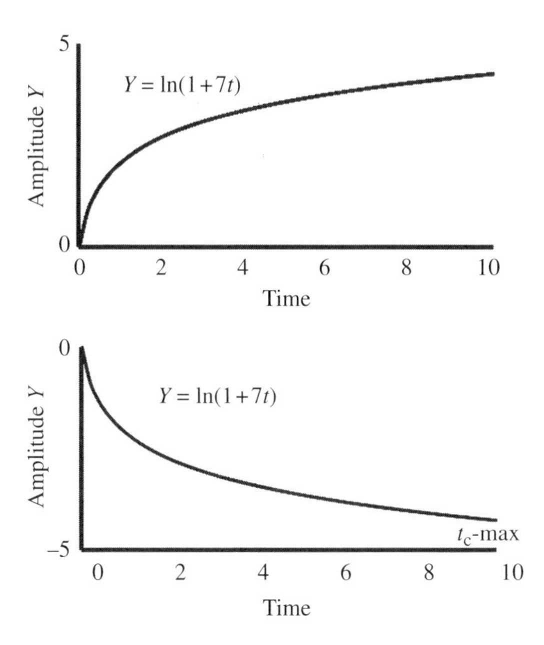

Figure 6.2 Examples of ln(1 + B time) aging law, with upper graph similar to primary and secondary creep stages and the lower graph similar to primary battery voltage loss

of parametric logarithmic-in-time aging can further be put in perspective as it can be tied to catastrophic lognormal failure rates. As to why there is a connection in particular to the catastrophic lognormal failure rate distribution is explained in detail in Section 9.2 [1, 2].

Logarithmic-in-time aging has a similar form to a power law when the exponent of the time is between 0 and 1. For example

$$Y = At^k \text{ for } 0 < k < 1 \tag{6.9}$$

and

$$Y = B\ln(1 + at) \tag{6.10}$$

can both model the same physical degradation process. As an example, consider when $A = 1.45$, $k = 0.5$ in Equation (6.9) and when $B = 2.2$, $a = 0.7$ in Equation (6.10). Results in Figure 6.3 show how these two equations overlap graphically in modeling appearance.

The point is that many degradation processes in the literature, such as secondary creep, use empirical power-law models instead of the logarithmic model. However, a logarithmic-in-time model may equally be suitable for which a TAT model could be found. In this case, we might have a better understanding of the degradation process when physics can be applied to the degradation process to help understand what is occurring. Applications for the TAT model are provided in the next chapter.

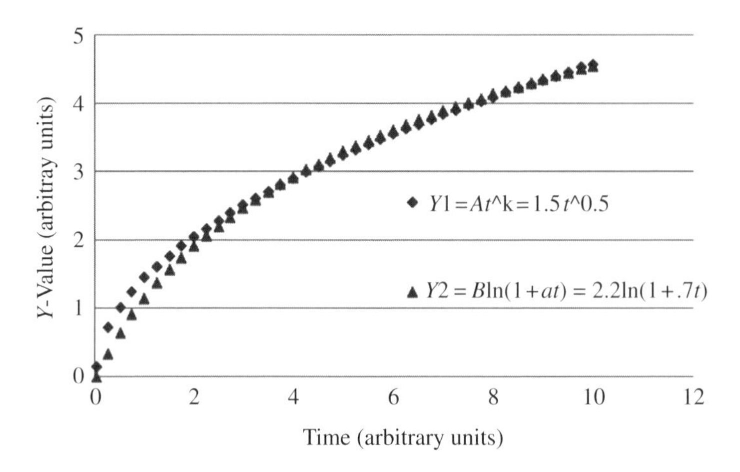

Figure 6.3 Log time compared to power law aging models

6.3 Free Energy Use in Parametric Degradation and the Partition Function

Here we wish to describe more information on the probability distribution used in Equation (6.2). In statistical mechanics the Boltzmann distribution gives the probability that a system will be in a certain state as a function of the state's energy and the temperature of the system.

In 1868 Ludwig Boltzmann formulated what is now a popular distribution for describing the probability for a system to be in a particular energy state as a function of the state's temperature. The distribution is given by

$$p_i = \frac{\exp(-\varepsilon_i/k_B T)}{\sum_i^N \exp(-\varepsilon_i/k_B T)} = \frac{\exp(-\varepsilon_i/k_B T)}{Z} \tag{6.11}$$

where the system has a probability p_i of being in the ith state; k_B is the Boltzmann constant; T is the temperature of the system; N is the number of possible states that the system can be in; and the denominator Z is known as the partition function.

Just as the entropy has a statistical definition, the Helmholtz free energy is also related to the partition function

$$F \approx -k_B T \ln Z, \tag{6.12}$$

or we can write, from Equations (6.11) and (6.12), that the partition function is

$$Z(T) = \sum_i^N \exp(-\varepsilon_i/k_B T) = \exp(-F(U,T)/k_B T). \tag{6.13}$$

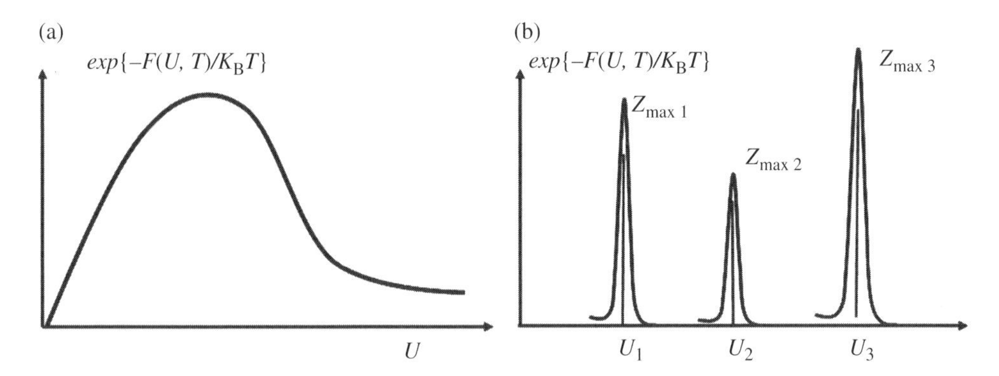

Figure 6.4 (a) Continuous function with numerous energy states. (b) Relative minimum energy states having different degradation mechanisms

The sum, often referred to as an ensample, can have a number of different internal energy states as shown in Figure 6.4a. The partition function near a true equilibrium state will have the free energy at a minimum; when the free energy is at a minimum, the value and the partition function will be at a maximum value. However, complex systems can have numerous degradation mechanisms each with their own relative minimum free energy equilibrium state, with relative maximum value for the partition function which for the ensemble of states are really the only contributing state in the sum that contribute to the value of Z as shown in Figure 6.4b. Therefore, the probability for a state to be occupied near an equilibrium state according to the ith relative minimum (where in Figure 6.4 $i = 1, 2,$ or 3) is

$$p_i = p_0 \exp\left(-\frac{F_i(U,T)}{k_B T}\right). \tag{6.14}$$

The activation energy of the ith failure mechanism, $E_{a,i}$, is related to the relative minimum barrier height (see Figure 6.1) which, using the partition function method and Figure 6.4b and according to Equation (6.12) is

$$E_{a,i} = F_i = -k_B T \ln Z_{max,i}. \tag{6.15}$$

The probability of a defect occurring then requires enough energy to surmount the free energy barrier created by the relative minimum in the free energy for the particular thermally activated failure mechanism, and is given by Equation (6.2):

$$\frac{dp}{dt} = \nu \exp\left(-\frac{\phi}{k_B T}\right).$$

We now can identify $\phi = F$ as the Helmholtz free energy, described in more detail by the activation energy through the Boltzmann distribution statistical mechanics. This identifies Equation (6.2) as useful in terms of a system's free energy near either a relative minimum or a true equilibrium thermodynamic state. Note that the free energy can also be written in terms of the Gibbs free energy (see Equation (5.42), for example).

Table 6.1 Failure mechanisms and associated thermal activation energies

Failure mechanism	Stress	Activation energy (eV)
Dielectric breakdown	Electric field, temperature	0.2–1.0
Corrosion	Temperature, humidity, voltage	0.3–1.1
Electromigration	Temperature, current density	0.5–1.2
Au–Al intermetallic growth	Temperature	1.0–1.05
Hot carrier injection	Electric field, temperature	0.9–1.1
Slow charge trapping	Electric field, temperature	1.0–1.3
Mobile ionic contamination	Temperature	1.0–1.05

Now if the degradation of a system is a thermally activated mechanism, it is likely to have a *relative* minimum in the free energy. This can occur whether one is concerned with soft degradation over time having a parametric failure mechanism or even a catastrophic failure mechanism. Table 6.1 lists a number of different failure mechanisms with their typical known historical activation energies which have been observed. The method for measuring the activation free energy barrier height is exemplified in the Special Topics B section of this book. Systems are typically at a relative minimum for most of the lifetime of the product and do not catastrophically fail until enough defects have accumulated by gaining enough energy to jump over the free energy barrier. As defects accumulate at some point, the system loses its strength and catastrophically fails. This is the point where the system's free energy has descended to its true minimum value which is its final equilibrium state, and the point where catastrophic failure occurs.

6.4 Parametric Aging at End of Life Due to the Arrhenius Mechanism: Large Parametric Change

In Section 6.2.1 we found aging due to small parametric change. Here we consider the second case for larger parametric change. A second TAT model can be obtained for both the initial aging period and end of life using both terms in the Maclaurin expansion in Equations (6.3) and (6.4) and performing the integration. The results obtained [1, 2] can be written:

$$a = \xi + b\,\mathrm{erf}^{-1}\left[\exp\left(-K^2\right)\left(\beta\frac{\nu(T)}{b}\right)t + \mathrm{erf}(K)\right] \tag{6.16}$$

where erf and erf^{-1} are the error function and its inverse, respectively, and

$$\xi = y_1/y_2,\ b = -2k_\mathrm{B}T/y2,\ K = \xi/b,\ \text{ and }\ \beta = 2/\sqrt{2\pi}.$$

This model is a parametric aging phenomenon that ages similar to logarithmic-in-time models and quickly goes catastrophic at the end of its life due to Arrhenius degradation. This is illustrated in Figure 6.5. The figure shows that aging starts off similar to logarithmic-in-time aging, but then quickly goes catastrophic at the critical corresponding time t_c [1, 2].

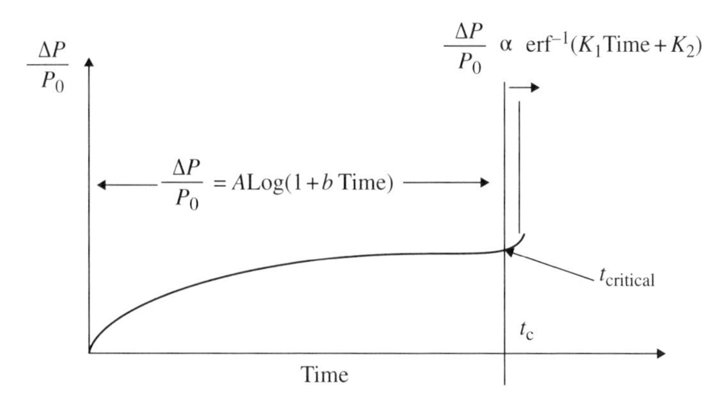

Figure 6.5 Aging with critical values t_c prior to catastrophic failure. Source: Feinberg and Widom [1]. Reproduced with permissions of IEEE

Figure 6.5 illustrates a number of rate processes that start off with log(time) aging then suddenly go catastrophic. Some examples exhibiting forms of this dependence over time are batteries [1], the three phases of creep, and cold-worked metals recrystallizing. What is interesting in this model is that the rate of initial aging is mathematically connected to its rate of final catastrophic behavior in this model. This suggests that, if the initial aging process is truly understood, a catastrophic prognostic may be possible.

Summary

6.1 Free Energy Roller Coaster

The activation energy can be thought of as the amount of energy needed for the degradation process. This defines a special relative equilibrium aging state

$$\phi_B \leq \phi \leq \phi_B + \Delta\phi \tag{6.1}$$

where the subscript B refers to the barrier and $\Delta\phi$ is the barrier height.

In the TAT model the aging process is closely related to the parametric change. The assumption is that the free energy itself will be associated with the parameter through the thermodynamic work.

6.2.1 Arrhenius Aging Due to Small Parametric Change

The TAT model describes logarithmic-in-time parametric aging

$$a = \frac{\Delta P}{P} \cong A \ln(1 + Bt) \quad \text{for } a << 1, \tag{6.7}$$

where A and B are defined

$$A = \frac{k_B T}{y_1} \quad \text{and} \quad B = \frac{\nu(T) y_1}{k_B T}. \tag{6.8}$$

Logarithmic-in-time aging is an important process since the origin of this aging kinetics can mathematically be tied to the Arrhenius mechanisms, of which numerous examples exist [1, 2]: aging and kinetic processes such as frequency aging drift [3]; corrosion of thin films; chemisorption processes [4]; early degradation of primary battery life [5]; activated creep (see Section 7.3); activated wear (see Section 7.2); transistor key parameter aging (see Section 7.4); and so forth. The significance of parametric logarithmic-in-time aging can further be put in perspective as it can be tied to catastrophic log-normal failure rates. Why there is a connection in particular to the catastrophic log-normal failure rate distribution is explained in detail in Section 9.2 [1, 2].

6.3 Free Energy Use in Parametric Degradation and the Partition Function

Just as the entropy has a statistical definition, the Helmholtz free energy is also related to the partition function

$$F \approx -k_B T \ln Z, \tag{6.12}$$

where Z is known as the partition function. Alternatively, from Equations (6.11) and (6.12) we can write the partition function as

$$Z(T) = \sum_i^N \exp(-\varepsilon_i / k_B T) = \exp(-F(U,T)/k_B T). \tag{6.13}$$

However, complex systems can have numerous degradation mechanisms, each with their own relative minimum free energy equilibrium state with relative maximum value for the partition function. For the ensemble of states, these are really the only contributors to the value of Z as shown in Figure 6.4b. Therefore, the probability of a state being occupied near an equilibrium statem according to the ith relative minimum (where in Figure 6.4 $i = 1, 2,$ or 3) is

$$p_i = p_0 \exp\left(-\frac{F_i(U,T)}{k_B T}\right). \tag{6.14}$$

The probability of a defect occurring then requires enough energy to surmount the free energy barrier created by the relative minimum in the free energy for the particular thermally activated failure mechanism, and is given by Equation (6.2):

$$\frac{dp}{dt} = \nu \exp\left(-\frac{\phi}{k_B T}\right).$$

We now can identify $\phi = F$ as the Helmholtz free energy, described in more detail by the activation energy through the Boltzmann distribution statistical mechanics. Note that the free energy can also be written in terms of the Gibbs free energy (see Equation (5.42) for example).

6.4 Parametric Aging at End of Life Due to the Arrhenius Mechanism

A second TAT model can be obtained for both the initial aging period and end of life using both terms in the Maclaurin expansion in Equations (6.4) and (6.5) and performing the integration, obtaining [1, 2]:

$$a = \xi + b\,\mathrm{erf}^{-1}\left[\exp\left(-K^2\right)\left(\beta\frac{\nu(T)}{b}\right)t + \mathrm{erf}(K)\right] \tag{6.16}$$

where erf and erf^{-1} are the error function and its inverse and

$$\xi = y_1/y_2,\ b = -2k_{\mathrm{B}}T/y2,\ K = \xi/b,\ \text{ and }\ \beta = 2/\sqrt{2\pi}.$$

This model is a parametric aging phenomenon that ages similar to logarithmic-in-time models and quickly goes catastrophic at its end of life due to Arrhenius degradation. This is illustrated in Figure 6.5.

References

[1] Feinberg, A. and Widom, A. (2000) On thermodynamic reliability engineering. *IEEE Transaction on Reliability*, **49** (2), 136.

[2] Feinberg, A. and Widom, A. (1996) Connecting parametric aging to catastrophic failure through thermodynamics. *IEEE Transaction on Reliability*, **45** (1), 28.

[3] Warner, A.W., Fraser, D.B. and Stockbridge, C.D. (1965) Fundamental studies of aging in quartz resonators. *IEEE Transaction on Sonics and Ultrasonics*, **12**, 52.

[4] Ho, Y.-S. (2006) Review of second-order models for adsorption systems. *Journal of Hazardous Materials*, **B136**, 681–689.

[5] Linden, D. (ed) (1980) *Handbook of Batteries and Fuel Cells*, McGraw-Hill, New York.

7

TAT Model Applications: Wear, Creep, and Transistor Aging

7.1 Solving Physics of Failure Problems with the TAT Model

In this chapter we provide examples of the thermally activated time-dependent (TAT) models that include degradation models for wear, creep, transistor aging, as well as dielectric leakage. We include both cases in the instant of field-effect transistor (FET) and bipolar transistors.

7.2 Example 7.1: Activation Wear

There are a number of different types of wear, often characterized by their wear rate [1, 2]. Activation wear is a term we use here to describe wear due to a thermally activated process. This approach offers an alternative to the Archard's empirical constant-type-wear model discussed in Chapter 4 (see Equation (4.20)). Heat is often an enemy in degradation processes and leads to an acceleration effect in frictional wear. When logarithmic-in-time aging occurs, as is illustrated in Figure 7.1 (often observed in metals), the TAT model can apply. This is characterized by an initially high wear rate followed by a steady-state low wear rate. Archard's type shows the case of steady wear in time, where the wear rate amount is constant over the sliding distance (Figure 7.1). Log(time) shows the case of an initially constant wear rate, where surface roughness increases to a certain value and does not increase much after that as the wear rate then decreases over sliding distance. The other type of wear is observed in lapping and polishing for surface finishing of ceramics.

Here we will focus on activation or log(time) wear and consider surface mass trapped in a potential well. We need to apply enough thermal energy (friction) for mass to escape from the chemical bonds. When a normal force is applied to the surface due to contact wear under a constant velocity, the normal force P results in material mass M removal (see Figure 4.3). We will start by taking a naïve kinetic friction approach

Thermodynamic Degradation Science: Physics of Failure, Accelerated Testing, Fatigue, and Reliability Applications, First Edition. Alec Feinberg.

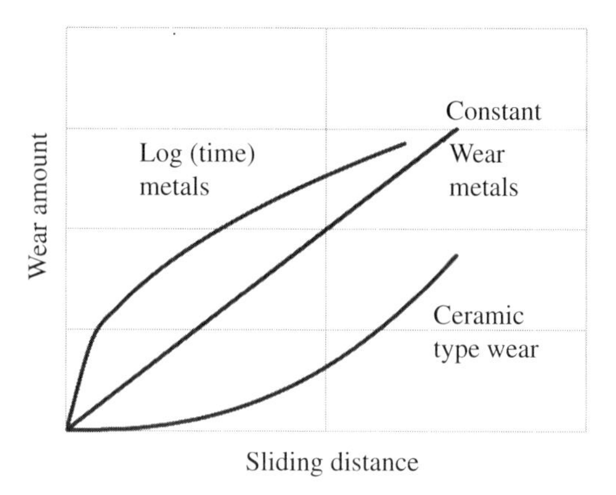

Figure 7.1 Types of wear dependence on sliding distance (time)

$$F_N = \frac{d}{dt}(M\nu) = \frac{dM}{dt}(\nu) \propto (\nu)\left(\mu_{\mathrm{K,eff}}P\right) \tag{7.1}$$

where $\mu_{\mathrm{K,eff}}$ is an effective coefficient of sliding kinetic friction and the velocity ν is constant. Here we start with a simple kinetic frictional model to get the general form. For an activated process of mass removal, we write that the change in the mass per unit time is proportional to the normal force applied to the contact surface so that

$$\frac{dM}{dt}\nu = P\mu_{\mathrm{K,eff}}(T) = P\mu_{\mathrm{K,eff}}\exp\left(-\frac{\phi(M)}{k_{\mathrm{B}}T}\right) \tag{7.2}$$

where $\mu_{\mathrm{K,eff}}(T)$ is a temperature-dependent effective kinetic coefficient of sliding friction which must be generalized. This can be written

$$\frac{dM}{dt} = \gamma\exp\left(-\frac{\phi(M)}{k_{\mathrm{B}}T}\right) \tag{7.3}$$

where $\gamma = \mu_{\mathrm{K,eff}}P/\nu$. We now have to depart from our naïve approach as kinetic friction, although insightful, is not helpful enough in accurately depicting the amplitude in wear. To be consistent and specify the model a bit further, we can use historical information such as the Archard's wear model in Equation (4.20). We have a few options and can write the amplitude

$$\gamma = \frac{kP\nu\rho}{H} = \frac{\bar{M}_0}{\tau} \tag{7.4}$$

That is, we can either make use of the Archard's amplitude or write the amplitude in experimental terms of an initial average mass removal $\bar{M}_0$ and an average time constant for removal τ. This amplitude can be found experimentally as we will describe. We have added ρ, the density of the softer material, to Equation (4.20) for consistent units with mass removal and the other parameters defined in this equation. Expanding the free energy as a function of the mass removal we have

$$\phi(M) \approx \phi(0) + \frac{d\phi}{dM} M + \ldots = \phi(0) + \mu M + \ldots \tag{7.5}$$

We identify $\mu = d\phi/dM$ (see Equation (2.114) or (5.7)) as the *activation chemical potential per unit mass*, which now gives

$$\frac{dM}{dt} = \gamma(T) = \gamma_0 \exp\left(-\frac{\phi(0) + \mu M}{k_B T}\right) = \gamma(T) \exp\left(-\frac{\mu M}{k_B T}\right) \tag{7.6}$$

and ϕ is the activation energy for the wear process that can be found experimentally by testing at different temperatures as explained below. Rearranging terms, and solving for the mass as a function of time t and integrating, provides a logarithmic-in-time aging TAT model in terms of mass removal over time:

$$M = A \ln(1 + Bt) \tag{7.7}$$

where A and B are

$$A = \frac{k_B T}{\mu} \quad \text{and} \quad B = \frac{\gamma(T)\mu}{k_B T}. \tag{7.8}$$

Note the logarithmic-in-time dependence in the activation wear case differs from Archard's linear dependence (i.e., $l = vt$). Here we find that when Bt in Equation (7.7) is less than or of the order of 1, the removal amount is large at first then is less as time accumulates (non-linear in time removal). However, when $Bt >> 1$, the removal is in $\ln(t)$ dependence. Also since $\ln(1 + X) \sim X$ for $X << 1$, then Equation (7.7) can be approximated as $M \approx ABt = \gamma(T)t$ for $Bt << 1$, which upon substitution of Equation (7.4) agrees with Equation (4.20), the Archard wear equation. Note that, one could make use of this early time approximation to help determine $\gamma_0 = \bar{M}_0/\tau$ and the wear activation energy ϕ experimentally.

7.3 Example 7.2: Activation Creep Model

Activation creep is a term we use here to describe creep due to a thermally activated process where the activation free energy is a function of the strain itself. This approach offers an alternative approach to modeling the creep process. The general model indicates that for creep to occur under temperature activation that a number of dislocations will occur over time in the metal lattice (see Figure 3.1). This when, due to temperature stress, dislocations N have a probability of hopping over a potential barrier and weakening the crystal metal lattice. Its hopping rate of occurrence is (see also Equation (6.2))

$$\frac{dN}{dt} = N_0 \exp\left(-\frac{\phi(N)}{k_B T}\right) \tag{7.9}$$

where in the TAT model the activation energy is a function of N. In the case of creep, N would be proportional to strain. However, this relation must of course be modified when a load σ is applied and the material is also subjected to mechanical means. We will therefore add this using a popular form of the empirical creep rate equation, typically written

$$\frac{d\varepsilon}{dt} = \varepsilon_0 \sigma^N \exp\left(-\frac{\phi(\varepsilon)}{k_B T}\right). \tag{7.10}$$

Note the difference from Equation (4.12) regarding the time dependence of this popular form for the creep rate. Also the key difference in this approach is that we associated the creep rate process with the activation free energy as a function of the strain itself. We will at this point use the TAT model by expanding it in terms of the strain dependence in a Maclaurin series:

$$\phi(\varepsilon) \approx \phi(0) + \frac{d\phi}{d\varepsilon}\varepsilon + \ldots \tag{7.11}$$

We identify that the change in the free energy is due to damage that occurred from mechanical thermodynamic conjugate work (see Table 1.1 and Chapter 4):

$$\frac{d\phi}{d\varepsilon} = \frac{\delta W}{d\varepsilon} = \sigma. \tag{7.12}$$

This then yields

$$\frac{d\varepsilon}{dt} = \varepsilon_0 \sigma^N \exp\left(-\frac{\phi(\varepsilon)}{k_B T}\right) = \varepsilon_0(T)\sigma^N \exp\left(-\frac{\sigma\varepsilon}{k_B T}\right) \tag{7.13}$$

which is now in the form of the TAT model (see Equations (6.7) and (6.8)). We find on integration:

$$\varepsilon = \frac{\Delta L}{L} \cong A \ln(1 + Bt) \quad \text{for } \varepsilon << 1 \tag{7.14}$$

where

$$A = \frac{k_B T}{\sigma} \quad \text{and} \quad B = \frac{\varepsilon(T)\sigma^{N+1}}{k_B T} \tag{7.15}$$

or

$$\varepsilon = \frac{\Delta L}{L} \cong \frac{k_B T}{\sigma} \ln\left[1 + \frac{\varepsilon(T)}{k_B T}\sigma^{N+1} t\right] \tag{7.16}$$

where

$$\varepsilon(T) = \varepsilon_o \exp\left(-\frac{\phi(0)}{k_B T}\right).$$

We note that we actually had to start with an empirical expression in Equation (7.10) for the creep rate. Once we expanded the activation free energy in terms of the strain, the creep rate indicated a non-linear time dependence $\ln(1 + Bt)$. The logarithmic-in-time dependence model has the common curvature found in primary and secondary creep stages (see Figures 9.5 and 9.6 for creep curves and compare to Figure 6.2). A power law dependence $\varepsilon \, \alpha \, t^{\beta}$ where $0 < \beta < 1$ is also popular as highlighted by Equation (4.12), and both the $\ln(1 + Bt)$ and power law form have been used by other authors [3]. (Also see Figure 6.3 on how a power law can be similar to log time aging rates.) Interestingly enough we see the stress amplitude is an inverse relation to strain, which of course by itself would be incorrect. However, the argument in the natural log function includes strain to the $N+1$ power which, as long as $N > 0$, will actually show that the strain increase with stress. Finally, the activation energy should be the same $\phi(0)$ as reported in the literature. The third stage of creep can also be modeled using the TAT model. This concept is provided at the end of this chapter.

7.4 Transistor Aging

Here we illustrate how we can extend the TAT model to transistor aging [4]. We are primarily concerned with key transistor device parameters. In the bipolar case for the common-emitter configuration, the key transistor parameter is beta aging, showing it to be directly proportional to the fractional change in the base-emitter leakage current. In the FET case, the key transistor parameter considered is transconductance aging that results from a change in the drain-source resistance and gate leakage current. Then the TAT model is used to provide an aging expression that accounts for the time degradation of these parameters found in life test. These expressions provide insight into degradation that links aging to junction-temperature-dependent mechanisms. The mechanisms for leakage can be thought of as similar to a corrosion process having a corrosion current. All the components are similar (an anode, cathode, a conducting path, and an effective type of dielectric "electrolyte"). Some typical life test data on heterojunction bipolar transistors (HBTs) and metal semiconductor field-effect transistor (MESFETs) are illustrated.

7.4.1 Bipolar Transistor Beta Aging Mechanism

Bipolar transistor degradation over time is a serious issue in microelectronics.

Generally, there are two common bipolar aging mechanisms: an increase in emitter ohmic contact resistance and an increase in base leakage currents.

Both can be thought of as corrosion-occurring process. Since β is given by I_{ce}/I_{be} in the common emitter configuration, any base leakage degradation in I_{be} will degrade β. In this section, we describe a model for β degradation [4] over time due to leakage. We start by considering a change in the base current gain for the common emitter configuration as

$$\beta(t) = \beta_0 - |\Delta\beta(t)| \tag{7.17}$$

where β_0 is the initial value (prior to aging) of I_{ce}/I_{be}. The time-dependent function $\Delta\beta(t)$ can be found through the time derivative:

$$\dot{\beta}(t)=\frac{d\beta}{dt}=\frac{d}{dt}\left(\frac{I_{ce}}{I_{be}}\right)=\frac{\dot{I}_{ce}}{I_{be}}-\frac{I_{ce}}{I_{be}^2}\dot{I}_{be}=\beta_0\left(\frac{\dot{I}_{ce}}{I_{ce}}-\frac{\dot{I}_{be}}{I_{be}}\right). \tag{7.18}$$

Approximating d/dt by $\Delta/\Delta t$, and noting that Δt is common to both sides of the equation and cancels, yields

$$\Delta\beta(t)\approx\beta_0\left(\frac{\Delta I_{ce}(t)}{I_{ce}}-\frac{\Delta I_{be}(t)}{I_{be}}\right)\approx-\beta_0\left(\frac{\Delta I_{be}(t)}{I_{be}}\right)=\beta_0\left(\frac{|\Delta I_{be}(t)|}{I_{be}}\right). \tag{7.19}$$

In the above equation, ΔI_{ce} has been set to zero as no change in this parameter is usually observed experimentally. We added the absolute value sign since beta overall will degrade $(\beta(t)=\beta_0-\Delta\beta(t))$ as the leakage current to the base increases. The first result is therefore that the change in β is directly proportional to the fractional change in the base-emitter leakage current:

$$\beta(t)=\beta_0\left\{1-\left|\frac{\Delta I_{be}(t)}{I_{be}}\right|\right\}=\beta_0\left\{1-\left|\frac{\Delta V_{be}(t)}{V_{be}}\right|\right\}. \tag{7.20}$$

The last expression for the voltage follows from the next section.

7.4.2 *Capacitor Leakage Model for Base Leakage Current*

At this point, base charge storage is discussed in order to develop a useful capacitive model. When a transistor is first turned on, electrons penetrate into the base bulk gradually. They reach the collector only after a certain delay time τ_d. The collector current then starts to increase in relation to the current diffusion rate. Concurrent with the increase of the collector current, is excess charges build up in the base. As a first approximation, the collector current and excess charge increases in an exponential manner with time constant τ_b. This transient represents the process of charging a “capacitor” in the simplest of RC circuits shown in Figure 7.2 and

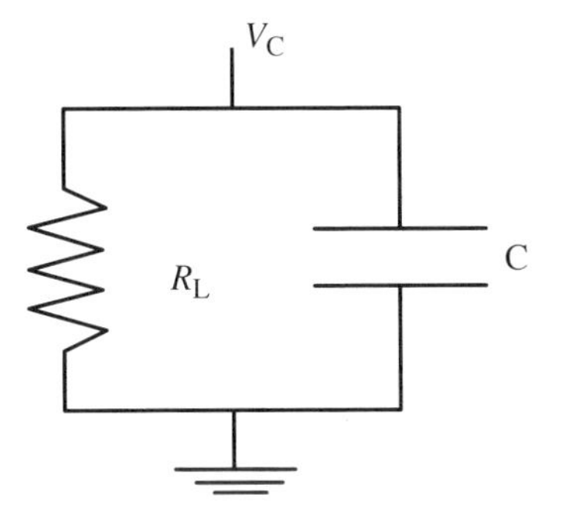

Figure 7.2 Capacitor leakage model

dissipating through the resistance R. We use this approximation to provide a simple model for base leakage. The steady-state value of excess charge build-up in base-emitter bulk Q_k is then

$$Q_k = (Q_{be})_k \cong (C_{be} V_{be})_k = (C_{be}(I_{be} R_{be}))_k = (I_{be} \tau_b)_k \tag{7.21}$$

where $\tau_b = R_{be} C_{be}$ is the time constant for steady-state excess charges in the base-emitter junction. As discussed above, the base-emitter junction primarily contributes to aging effects. Along with this bulk effect is parasitic surface charging Q_s and leakage. We can also treat these using a simple RC charging model. In this view, the surface leakage can be expressed as

$$Q_s = (Q_{be})_s \cong (C_{be} V_{be})_s = (C_{be}(I_{be} R_{be}))_s = (I_{be} \tau_b)_s. \tag{7.22}$$

The total excess charging at the base is due to surface and bulk leakage

$$Q_{be} = Q_s + Q_k. \tag{7.23}$$

This indicates that as the transistor ages, Q_{be} increases along with I_b. Some of the increase in Q_{be} is caused by the increase in impurities and defects in the base surface and bulk regions due to operating stress.

The impurities and defects cause an increase in electron scattering and an increase in the probability for trapping and charging and eventual recombination in the base. The above feature leads to an increased leakage current. In the capacitive model shown in Figure 7.2, incremental changes are

$$dQ = C\,dV = C\,R\,dI = \tau dI$$

where we view Q, V, and I as time-varying with age, that is,

$$\Delta\beta(t) \cong \beta_0 \left(\frac{|\Delta I_{be}(t)|}{I_{be}} \right) = \beta_0 \left(\frac{|\Delta Q_{be}(t)|}{Q_{be}} \right) = \beta_0 \left(\frac{|\Delta V_{be}(t)|}{V_{be}} \right). \tag{7.24}$$

Thus, a second result is that the change in β is proportional to the fractional change in the base-emitter leakage current, charge, and voltage.

Experimentally, β degradation is observed to follow a logarithmic-in-time aging as exemplified in Figure 7.3. Data plotted on C-doped MBEHBT device at 235°C at 10 kA/cm^2 are shown in Figure 7.3 [4]:

$$\frac{\Delta\beta(t)}{\beta_0} = \frac{\Delta I_{be}(t)}{I_{be}} = \frac{\Delta q_{be}(t)}{q_{be}} = -A \log(1 + Bt). \tag{7.25}$$

7.4.3 *Thermally Activated Time-Dependent Model for Transistors and Dielectric Leakage*

The leakage current often leading to dielectric breakdown has historically been explored and there are numerous mechanisms, or a combination of several, providing a physical explanation of the origin of the current that flows through the dielectric. These models stem from the fact

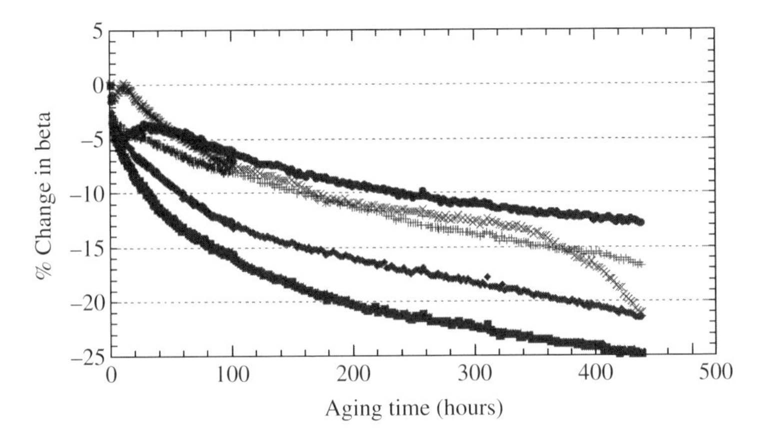

Figure 7.3 Beta degradation on life test data. Source: Feinberg *et al.* [4]. Reproduced with permission of IEST

that the work required to create defects that increase leakage current is thermally activated, and there is a reduction in the free energy barrier for defect generation due to the electric field lowering the barrier for defect creation. For example, in a bond breakage model, the E-field relates to the bonds breakage through the dipole energy. The energy barrier for creating defects has the form

$$\phi(q) \approx \phi(0) + \frac{d\phi}{dq} q + \dots \tag{7.26}$$

We identify that the change in the free energy is due to damage that occurred from electrical thermodynamic conjugate work (see Table 1.1):

$$\frac{\delta\phi}{dq} = \frac{\delta W}{dq} = V. \tag{7.27}$$

Common dielectric leakage mechanisms that relate to various forms of the free energy with different explanations include the: Poole–Frenkel effect; Schottky effect; thermo-chemical E-model; tunneling; and Fowler–Nordheim tunneling. The most general expression for current density j for these models is

$$j = CE^K \exp\left\{\frac{-(\phi_0 - \alpha E^N)}{k_B T}\right\} \tag{7.28}$$

where E is the electric field and α and C are constants. For the Poole–Frenkel [5] model $K = 1$, $N = 1/2$; for the Schottky effect [6] model $K = 0$, $N = 1/2$; for the thermo-chemical model $K = 0$, $N = 1$; and for the tunneling models $K = 2$, $N = -1$ [6].

We can start our leakage TAT model consistent with Equation (7.28) and use Equations (7.26) and (7.27) for the free energy with $N = 1$ and the general constant K as

$$\frac{dq}{dt} = q_0 V^K \exp\left(-\frac{(\phi(0) + qV)}{k_B T}\right) = q_0(T) V^K \exp\left(-\frac{qV}{k_B T}\right). \tag{7.29}$$

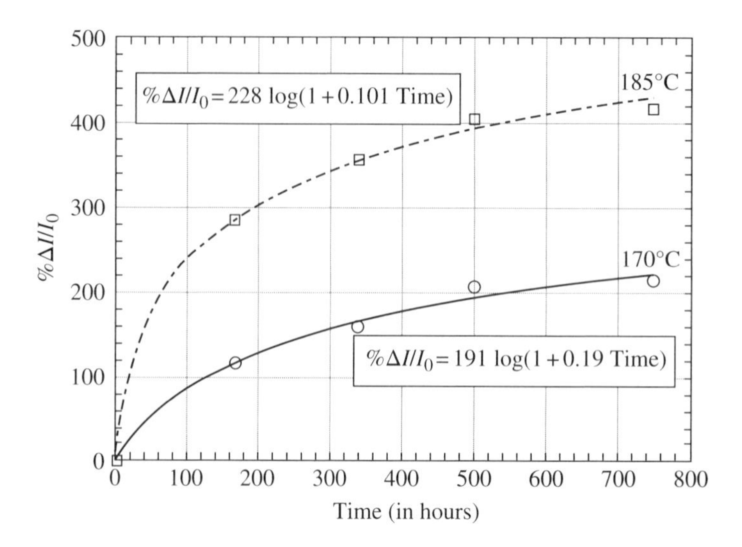

Figure 7.4 Life test data of gate-source MESFET leakage current over time fitted to the ln(1 + *Bt*) aging model. Junction rise was about 30°C. Source: Feinberg *et al.* [4]. Reproduced with permission of IEST

This is in the form of the TAT model results (see Equations (6.7) and (6.8)) where we find

$$\Delta\beta = \frac{\Delta q_{\text{be}}}{q_{\text{be}}} = \frac{\Delta I_{\text{be}}}{I_{\text{be}}} \cong A\ln(1+Bt) \quad \text{for } \Delta\beta_{\text{loss}} << 1 \tag{7.30}$$

where

$$A = \frac{k_{\text{B}}T}{I_{\text{be}}V} \quad \text{and} \quad B = \frac{q(T)V^{K+1}}{k_{\text{B}}T}.$$

Alternatively,

$$\Delta\beta(t) \cong \frac{k_{\text{B}}T}{VI_{\text{be}}}\ln\left[1+\frac{q(T)}{k_{\text{B}}T}V^{K+1}t\right] \tag{7.31}$$

and

$$q(T) = q_0\exp\left(-\frac{\phi(0)}{k_{\text{B}}T}\right).$$

We expect $K > 0$ for Beta degradation $\beta(t) = \beta_0 - |\Delta\beta(t)|$ to increase with $V(=Ed)$. The result of this logarithmic-in-time model yields a good fit to the beta degradation data in Figure 7.3 and for the FET data in Figure 7.4.

7.4.4 Field-Effect Transistor Parameter Degradation

In this section, transconductance degradation over time is described to help understand aging in FET devices such as MESFETs. Unlike the beta parameter for the bipolar case, we note that transconductance can also apply to bipolar or FET devices. This methodology applied to FETs can also be extended to the bipolar case. The transconductance in the FET case is of interest as a small change in the gate-source voltage V_{gs} can make a large change in the current flowing out of the drain I_{ds} of the device. The transconductance g_m relates these two variables as $I_{ds} = g_m(t) V_{gs}$. It is this amplification factor that we will now be concerned about for degradation issues.

A key issue similar to bipolar transistor degradation in FET type devices is again leakage current [4].

We start by modeling a change in the transconductance g_m similar to beta change as

$$g_m(t) = g_0 - |\Delta g_m(t)| \tag{7.32}$$

where the initial value g_0 is taken from the linear portion of the transconductance curve, that is

$$g_0 = \frac{I_{ds}}{V_{gs} - V_0}. \tag{7.33}$$

Here, we use the linear portion of the curve for simplicity. Similar results will follow for other portions of the curve. The time-dependent function $\Delta g_m(t)$ is found from its derivative as:

$$\dot{g}_m(t) = \frac{dg_m}{dt} = \frac{d}{dt}\left(\frac{I_{ds}}{V_{gs} - V_0}\right) = \frac{\dot{I}_{ds}}{V_{gs} - V_0} - \frac{Ids}{\left(V_{gs} - V\right)^2}\dot{V}_{gs} \tag{7.34}$$

or

$$\dot{g}_m(t) = \frac{I_{ds}}{V_{gs} - V_0}\left(\frac{\dot{I}_{ds}}{I_{ds}} - \frac{\dot{V}_{gs}}{V_{gs} - V_0}\right) = g_0\left(\frac{\dot{I}_{ds}}{I_{ds}} - \frac{\dot{V}_{gs}}{V_{gs} - V_0}\right). \tag{7.35}$$

We assume that the drain-source current change occurs as $dI/dt \sim (V/R^2)(dR/dt)$ with V_{ds} constant and voltage-gate change as $dV_{gs}/dt \sim d/dt(IR) = RdI_{gs}/dt$. Approximating d/dt by $\Delta/\Delta t$ and noting that Δt is common to both sides of the equation and canceling, the expression simplifies to

$$\Delta g_m(t) = g_0\left(\frac{\Delta R_{ds}}{R_{ds}} - \frac{\Delta I_{gs}}{I_{gs} - I_{gs,0}}\right). \tag{7.36}$$

Thus, the primary result for FETs is that transconductance aging arises from a change in the drain-source resistance and gate leakage. However, it is commonly found that resistance aging dominates the reaction [4]. As far as R_{ds} is concerned, resistance is related to scattering inside the drain-source channel $\Delta R_{ds}/R_{ds} = \Delta\rho_{ds}/\rho_{ds} = \Delta l_{ds}/l_{ds}$, where ρ is the resistivity, and l is the average mean-free path traveled by the electrons in the channel between collisions. This distance decreases as aging occurs and more defects occur in the channel, causing increased scattering.

At this point, we wish to point out that MESFET gate leakage aging data as shown commonly follows a logarithmic-in-time aging form similar to β degradation (a mechanism that we have modeled as dominated by leakage). This aging time dependence is illustrated in the life test data in Figure 7.4 over two aging temperatures [4]. The TAT model for gate leakage, similar to Equation (7.31) is

$$\frac{\Delta I_{gs}(t)}{I_{gs}} \cong -\frac{k_B T}{V I_{gs}} \ln\left(1 + I(T)\frac{V^{K+1}}{k_B T} t\right) \tag{7.37}$$

where

$$I(T) = I_0 \exp\left(-\frac{\phi(0)}{k_B T}\right).$$

This model is found to nicely fit the life test data in Figure 7.4.

Summary

7.2 Example 7.1: Activation Wear

Heat is often an enemy and wear can be viewed as a thermally activated process with frictional heating increasing the probability for mass removal. Wear has been observed in Type II metal processes to show logarithmic-in-time aging, for which we applied a TAT model in terms of mass removal over time

$$M = A \ln(1 + Bt) \tag{7.7}$$

where A and B are

$$A = \frac{k_B T}{\mu} \quad \text{and} \quad B = \frac{\gamma(T)\mu}{k_B T}. \tag{7.8}$$

When $Bt >> 1$, the removal is in ln(t) dependence and Equation (7.7) can be approximated as $M \approx ABt = \gamma(T)t$. For $Bt << 1$, when substituted into Equation (7.4) agrees with Equation (4.20), the Archard wear equation.

7.3 Example 7.2: Activation Creep Model

The entire creep curve can be shown to follow the TAT model. Specifically for stages 1 and 2 of creep, we find the following form of the TAT model:

$$\varepsilon = \frac{\Delta L}{L} \cong \frac{k_B T}{\sigma} \ln\left[1 + \frac{\varepsilon(T)}{k_B T}\sigma^{N+1} t\right] \tag{7.16}$$

where

$$\varepsilon(T) = \varepsilon_0 \exp\left(-\frac{\phi(0)}{k_B T}\right).$$

7.4.1 Bipolar Transistor Beta Aging Mechanism

Bipolar transistor degradation over time is a serious issue in microelectronics.

Generally, there are two common bipolar aging mechanisms: an increase in emitter ohmic contact resistance and an increase in base leakage currents.

Since transistor β gain is given by I_{ce}/I_{be} in the common-emitter configuration, any base leakage degradation in I_{be} will degrade β. Experimentally, β degradation is observed to fol-lows a logarithmic-in-time aging and the base-emitter junction primarily contributes to aging effects, as well as parasitic surface charging Q_s and leakage. We can treat these using a simple RC charging model as shown in Figure 7.2. This transient represents the process of charging a "capacitor" in the simplest of RC circuits shown in Figure 7.2 and dissipating through the resistance R. In this view, the surface and bulk leakage can be expressed as

$$Q_k = (Q_{be})_k \cong (C_{be}V_{be})_k = (C_{be}(I_{be}R_{be}))_k = (I_{be}\tau_b)_k \tag{7.21}$$

where $\tau_b = R_{be}C_{be}$.

The total excess charging at the base is due to surface and bulk leakage currents, that is:

$$Q_{be} = Q_s + Q_k. \tag{7.23}$$

As the transistor ages, Q_{be} increases with I_{be}. Some of the increase in Q_{be} is caused by the increase in impurities and defects in the base surface and bulk regions due to operating stress.

7.4.3 Thermally Activated Time-Dependent Model for Transistors and Dielectric Leakage

The TAT model found when we combine the simple RC leakage model for transistor beta degradation is

$$\Delta\beta(t) \cong \frac{k_B T}{V I_{be}} \ln\left[1 + \frac{q(T)}{k_B T} V^{K+1} t\right] \tag{7.31}$$

where

$$q(T) = q_0 \exp\left(-\frac{\phi(0)}{k_B T}\right).$$

We expect $K > 0$ for beta degradation $\beta(t) = \beta_0 - |\Delta\beta(t)|$ to increase with $V(=Ed)$. The result of this logarithmic-in-time model yields a good fit to the beta degradation data in Figure 7.3 and for the FET data in Figure 7.4.

7.4.4 Field-Effect Transistor Parameter Degradation

In FETs aging is due to transconductance degradation g_m over time. Results were obtained and shown to describe aging in MESFETs.

A key issue similar to bipolar transistor degradation in FET-type devices is again leakage current [4].

Transconductance change was shown to have the following results

$$\Delta g_m(t) = g_0\left(\frac{\Delta R_{ds}}{R_{ds}} - \frac{\Delta I_{gs}}{I_{gs} - I_{gs,0}}\right). \tag{7.36}$$

Thus, the primary result for FETs is that transconductance aging arises from a change in the drain-source resistance and gate leakage. However, it is commonly found that resistance aging dominates the reaction [4]. As far as R_{ds} is concerned, resistance is related to scattering inside the drain-source channel $\Delta R_{ds}/R_{ds} = \Delta\rho_{ds}/\rho_{ds} = \Delta l_{ds}/l_{ds}$, where ρ is the resistivity and l is the average mean-free path traveled by the electrons in the channel between collisions.

In terms of IDS gate leakage, MESFET data is shown to follow a logarithmic-in-time aging form. This time dependence is illustrated in Figure 7.4. The TAT model obtained here for gate leakage similar to Equation (7.31) is

$$\frac{\Delta I_{gs}(t)}{I_{gs}} \cong -\frac{k_B T}{V I_{gs}} \ln\left(1 + I(T)\frac{V^{K+1}}{k_B T}t\right) \tag{7.37}$$

where

$$I(T) = I_0 \exp\left(-\frac{\phi(0)}{k_B T}\right). \tag{7.38}$$

MESFET data followed this aging nicely.

References

[1] Kato, K. and Adachi, K. (2001) Wear mechanisms, in *Modern Tribology Handbook* (ed B. Bhushan), CRC Press, Boca Raton.

[2] Chiou, Y.C., Kato, K. and Kayaba, T. (1985) Effect of normal stiffness in loading system on wear of carbon steel—part 1: severe-mild wear transition. *Journal of Tribology*, **107**, 491–495.

[3] Lubliner, J. (2008see Chapter 2) *Plasticity Theory*, Dover Publications, Mineola.

[4] Feinberg, A., Ersland, P., Kaper, V. and Widom, A. (2000) On aging of key transistor device parameters. *Proceedings: Institute of Environmental Sciences and Technology*, **46**, 231.

[5] Harrell, W.R. and Frey, J. (1999) Observation of Poole–Frenkel effect saturation in SiO_2 and other insulating films. *Thin Solid Films*, **352**, 195–204.

[6] Sze, S.M. (1981) *Physics of Semiconductor Devices*, John Wiley & Sons, Inc., New York.

8

Diffusion

8.1 The Diffusion Process

Diffusion is one of the four main types of aging described in Chapter 1. Diffusion often occurs in mass-transport processes, which occur essentially from three processes: (1) convection and stirring; (2) electrical migration due to an electric field; and (3) diffusion from a concentration gradient. The first two are categorized as being under the control of an external force. This includes subtle process such as corrosion and where mass is transported in the electrolyte solution to and from the electrodes. Of these three processes, diffusion is more of a spontaneous process where external work is not involved. Diffusion can also occur in corrosion and can be a rate-controlling step. This is often the case in "hot corrosion" or "aqueous corrosion" due to oxidation. Furthermore, many aging processes due to diffusion do not involve electrochemical transitions.

For a system that is diffusing into an environment due to a concentration gradient, its energy is related to the chemical potential change. The following equilibrium thermodynamic example clarifies the important role of the chemical potential.

8.2 Example 8.1: Describing Diffusion Using Equilibrium Thermodynamics

Consider a system that can exchange energy with its environment via the diffusion process. This is shown in Figure 8.1 which as n system particles of any type in contact with an environment consisting of n_{env} particles of another or the same type. The particles could be atoms, ion, or molecules representing a chemical species of interest.

Thermodynamic Degradation Science: Physics of Failure, Accelerated Testing, Fatigue, and Reliability Applications, First Edition. Alec Feinberg.

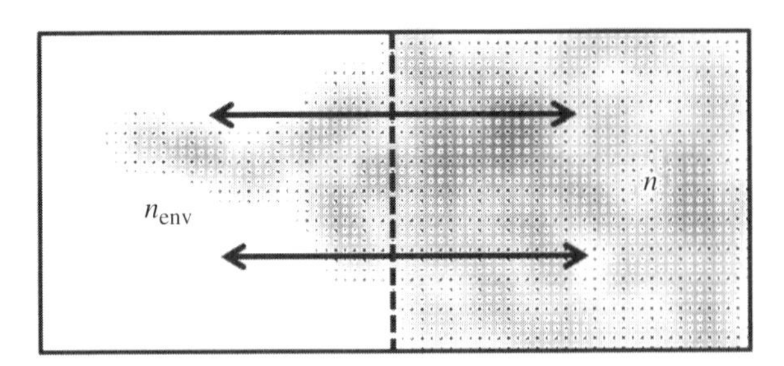

Figure 8.1 System with n particles and n_{env} environment particles

We can make an analogy to Example 2.10. Following a similar treatment, the spontaneous diffusion process seeks to maximize

$$S_{total} = S(U,n) + S_{env}(U_{env}, n_{env}). \tag{8.1}$$

The subscript env is for "environment" and we are not using a subscript here for the system. The system can exchange energy and atoms with the environment subject to the constraint of energy and particle conservation. The total conserved energy is

$$U_{total} = U + U_{env}, \tag{8.2}$$

and particle number is

$$n_{total} = n + n_{env}. \tag{8.3}$$

For a closed process this obeys

$$dU_{total} = dU + dU_{env} = 0, \tag{8.4}$$

$$dn_{total} = dn + dn_{env} = 0. \tag{8.5}$$

The entropy for the system and environment requires

$$dS_{total} = dS + dS_{env}. \tag{8.6}$$

Then we can write from the combined first and second law equation using the work conjugate pair for the chemical potential in Table 1.1 (also see Section 5.2):

$$dS_{total} = \left(\frac{dU}{T} - \frac{\mu dn}{T}\right) + \left(\frac{dU_{env}}{T_{env}} - \frac{\mu_{env} dn_{env}}{T_{env}}\right), \tag{8.7}$$

or, equivalently from Equations (8.4) and (8.5), $dn = -dn_{env}$ and $dU = -dU_{env}$ yielding

$$dS_{total} = \left(\frac{1}{T} - \frac{1}{T_{env}}\right) dU + \left(\frac{\mu_{env}}{T_{env}} - \frac{\mu}{T}\right) dn, \tag{8.8}$$

$$dS_{total} \geq 0. \tag{8.9}$$

In order to ensure that the total entropy goes to a maximum, we have positive increments under the exchange of energy and matter in which the system energy change is dU. If the system's chemical potential is greater than the environment's chemical potential, that is,$\mu > \mu_{\text{env}}$, then this dictates that $dn < 0$ and particles are diffusing out of the system side to the environment. If the system's chemical potential is less than the environment's chemical potential, that is, $\mu < \mu_{\text{env}}$, then this dictates that $dn > 0$. The particles are increasing in the system. The particles will then flow from the higher-chemical-potential region to the lower-chemical-potential region, similar to the potential of a battery.

If entropy is at the maximum value, then the first-order differential equation vanishes. Therefore, for equilibrium when the entropy is maximum $T = T_{\text{env}}$ and $\mu = \mu_{\text{env}}$, so that $dS_{\text{total}} = 0$. As well the pressure ($P = nRT/V$) is the same in the system and environmental regions at equilibrium in Figure 8.1 as the particles mix. Particles therefore migrate in order to remove differences in chemical potential. Diffusion ceases at equilibrium when $\mu = \mu_{\text{env}}$. The statement that diffusion occurs from an area of high chemical potential to low chemical potential is the same as saying that the diffusion process occurs from an area of high concentration to an area of low concentration. While this diffusion flow is a natural spontaneous process, the opposite cannot occur. That is, in accordance with the entropy maximum principle in Chapter 1, we see that the reveres process cannot occur without some work. The particles will not separate back once they are mixed without some amount of work done to accomplish this.

In a manner of speaking, the diffusion process has degraded so that no spontaneous energy exchange occurs and the system is in its lowest free energy state or maximum entropy state with the environment. Of course, we can have a quasistatic equilibrium condition when the diffusion process is fully concentrated and negligible particle exchange is occurring.

8.3 Describing Diffusion Using Probability

Diffusion can also be understood using a probability approach. Consider particles such as impurities; these impurities distribute themselves in space with passing time. For example, in semiconductors impurities deposited in optimal regions in space later diffuse to undesirable regions as the semiconductor ages. Raising the temperature may accelerate such aging.

To describe diffusion using probability, the Central Limit Theorem (Figure 8.2) is sometimes useful (see Special Topics A6). For example, the theorem applies for systems subject to a large number of small independent random effects as in a *random walk* (see Section 2.8.1). The central limit theorem is used in the sense that if we have X_1, X_2,..., X_N identically

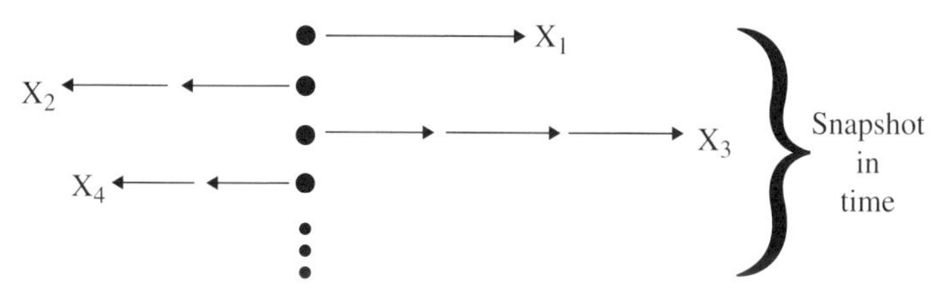

Figure 8.2 Diffusion concept

distributed random clusters of particles, each with a mean and variance, then the average of the entire system will approach a normal distribution as N approaches a large quantity of particles, moving as in a random walk manner. Here impurity particles are concentrated in a small region, each with an irregular random walk motion. From the Central Limit Theorem, the positions will become normally distributed in space for times short on a macroscopic scale but long on a microscopic scale. In one dimension, the distribution after time t will appear to be Gaussian.

Therefore, the probability P of finding a particle a distance x from the point of initial highest concentration taken as the origin, where one can center the mean (as in Figure 8.2), is

$$P(x) = \frac{1}{\sigma\sqrt{2\pi}} \exp\left[-\frac{1}{2}\left(\frac{x}{\sigma}\right)^2\right]. \tag{8.10}$$

In diffusion theory for this typical physical situation it is found that the particles spread linearly with time (t) with the variance as

$$\sigma^2 \propto t. \tag{8.11}$$

The proportionality constant is the diffusion coefficient D times 2 [1]:

$$\sigma^2 = 2Dt. \tag{8.12}$$

(Note that σ has the same units as meter and D has units of square meter per second.)

The probability of finding a particle at position x from the origin at time t in one-dimension is then

$$P(x, t, T) = \frac{1}{\sqrt{4\pi Dt}} \exp\left(-\frac{x^2}{4Dt}\right). \tag{8.13}$$

In terms of our semiconductor problem, if Q were the number of impurity particles in a unit area and C is the concentration of these impurities in the volume, the concentration distribution can be written

$$C(x, t) = \frac{Q}{\sqrt{4\pi Dt}} \exp\left(-\frac{x^2}{4Dt}\right). \tag{8.14}$$

The result is the solution to the diffusion equation with the boundary conditions for a physical situation described above. In one dimension, the diffusion equation that this satisfies is

$$\frac{\partial}{\partial t} C(X, t) = D(t) \frac{\partial^2}{\partial X^2} C(X, t). \tag{8.15}$$

The reader may be interested to show that the solution above satisfies this diffusion equation by direct substitution of Equation (8.14) into (8.15). It is important to note that the solution obtained is subject to the correct initial conditions.

8.4 Diffusion Acceleration Factor with and without Temperature Dependence

The diffusion coefficient itself is found to have Arrhenius temperature dependence

$$D(T) = D_0 \exp\left\{-\frac{\Delta}{RT}\right\} \tag{8.16}$$

where Δ is the barrier height (see Section 6.1). As T increases so does D (with an Arrhenius form). Note that since D occurs only as a product Dt (Equation (8.12)), the time scale is effectively changed (accelerating time) with an Arrhenius temperature dependence.

The diffusion acceleration factor will vary according to the diffusion rate ratio

$$\mathrm{AF}_D = \frac{\text{Diffusion rate}_{\text{stress}}}{\text{Diffusion rate}_{\text{use}}} = \mathrm{AF}_T \mathrm{AF}_x \tag{8.17}$$

where AF_T is given earlier in Equation (5.20) and AF_x is the acceleration factor due to spatial concentration gradient, defined:

$$\mathrm{AF}_x = \frac{\left(\dfrac{\partial^2 C(x,t)}{\partial x^2}\right)_{\text{stress}}}{\left(\dfrac{\partial^2 C(x,t)}{\partial x^2}\right)_{\text{use}}}. \tag{8.18}$$

8.5 Diffusion Entropy Damage

When describing diffusion using probability theory, we noted that the variance is a key intensive variable. Therefore, we anticipate entropy damage to be dominated by the variance. Diffusion is a continuous intensive variable of the system. In terms of degradation, we will consider that an increase in diffusion degrades the system of interest over time.

In Chapter 2 we noted that the entropy of a continuous variable is treated in thermodynamics using the concept of differential entropy which, for continuous variables as in the diffusion case, is

$$\text{Continuous}\, X, f(x):\ S(X) = -\int f(x) \log_2(f(x))dx = -E[\log f(x)]. \tag{8.19}$$

Now we have treated the particle distribution in the diffusion process as Gaussian. We then found from this function that the differential entropy results were given by (see Chapter 2, Equation (2.35))

$$S(X) = \frac{1}{2}\log\left[2\pi e \sigma(x)^2\right]. \tag{8.20}$$

This key result can be applied to the diffusion process here. We see that for entropy for a Gaussian diffusion system, the differential entropy is only a function of its variance σ^2 (independent from its mean μ). For a system that is diffusing over time, the entropy damage can then be measured in a number of ways where the change in the entropy at two different times t_2 and t_1 is

$$\Delta S_{\text{damage}} = S_{t2}(X) - S_{t1}(X) = \Delta S(t_2, t_1) = \frac{1}{2}\log\left(\frac{\sigma_{t2}^2}{\sigma_{t1}^2}\right). \tag{8.21}$$

Substituting from Equation (8.12), $\sigma^2 = 2Dt$, we write

$$\Delta S_{\text{damage}} = \frac{1}{2}\log\left(\frac{t_2}{t_1}\right) \tag{8.22}$$

where we assume $D_2 = D_1$. Therefore, entropy diffusion damage at a constant temperature, due to a concentration spatial gradient that is Gaussian, apparently occurs as logarithmic in time in this view.

This is an important result and actually agrees with previous theory that was presented in the thermally activated time-dependent (TAT) model (Chapters 6 and 7). There we found that a number of processes were described as aging in log time for thermally activated processes. Many natural aging processes have been observed to age with log(time) such as transistor degradation, crystal frequency drift, and early stages of creep. In fact, diffusion can play a role in degradation such as transistor aging (junction doping issues), intermetallic growth, contamination issues, and so forth. This result provides insight for the TAT model and these physical natural aging processes. Many complex aging processes can have a rate-controlling diffusion mechanism.

8.5.1 *Example 8.2: Package Moisture Diffusion*

A +85°C and 85% relative humidity (%RH) test is performed on a plastic molded semiconductor device. It is of interest to estimate how long it takes for the moisture to penetrate the mold and reach the die. We can estimate this time using the diffusion expression above. We use the experimentally reported values of:

$$\frac{\sigma^2}{2} = 0.85L^2 \tag{8.23}$$

where L is the molding compound thickness of 0.05 inches; $D_0 = 4.7\times10^{-5}\text{m}^2/\text{s}$ for moisture penetration into the mold; and $\Phi = 3\times10^{26}\text{eV/mole}$.

Solving the diffusion expression above (Equations (8.12) and (8.16)) with the time-dependent variance for t yields

$$t = \frac{\sigma^2}{2D} = \frac{0.85L^2}{D_0}\exp\left(\frac{\Phi}{RT}\right). \tag{8.24}$$

As a side note, this process does not appear at first to be a logarithmic-in-time model as might be suggested by Equation (8.22). However, we could first rearrange the above equation in terms of the following time-dependent power law

$$L=\sqrt{\frac{0.85}{D_0}\exp\left(\frac{\Phi}{2RT}\right)}\sqrt{t}. \tag{8.25}$$

We have noted in Section 6.2.1 (see Figure 6.3) that power laws of the form t^k where $0<k<1$ can be modeled with a log(time) form (see Equations (6.9) and (6.10)). That is, through proper modeling we anticipate that a TAT model can be found for this thermally activated process, in the form

$$L\approx A(T)\ln(1+at), \tag{8.26}$$

that would equally model the physics of the situation.

Now Equation (8.24) is in terms of R, the gas constant. It is instructive and simplest to put the exponential function in terms of Boltzmann's constant. Boltzmann's constant is by definition $k_B = R/N_A$. To put the expression in terms of Boltzmann's constant, one would divide through the exponential expression by Avogadro's number ($N_A = 6\times 10^{26}$ molecules/mole).

$$\exp\left(\frac{\Phi/N_A}{R/N_AT}\right)=\exp\left(\frac{E_a}{k_BT}\right) \tag{8.27}$$

Now $E_a = \Phi/N_A = 0.5$ eV. Inserting the numbers into the above expression yields

$$t=\frac{0.85(0.05\text{ inch}\times 0.0254\text{ m/inch})^2}{4.7\times 10^{-5}\text{ m/s}}\exp\left(\frac{0.5\text{ eV}}{(8.617\times 10^{-5}\text{eV/K})(85+273.16\text{ K})}\right) \tag{8.28}$$

$t=0.029$ s $\exp(16.2)=314\ 752$ s $=87$ h.

8.6 General Form of the Diffusion Equation

The most generalized diffusion equation for aging circumstances can include external forces, such as an electric field. For example, if the flux is a charged species and is driven by a force such as a constant electric field E, then [1]

$$\frac{\partial C(x,t)}{\partial t}=D\frac{\partial^2 C(x,t)}{\partial x^2}-\mu E\frac{\partial C(x,t)}{\partial x}. \tag{8.29}$$

Note that this equation shows that the diffusion equation can describe all three processes that we have categorized as fundamental to aging in Section 1.2: a thermally activated Arrhenius process with $D(T)$ dependence; the existence of a spatial gradation driving diffusion; and an external forced process. All processes are fundamentally driven by the thermodynamic state. The equation would be extremely difficult to solve if all mechanisms were equally important. However, aging can often be separated into its rate-controlling processes.

Summary

8.2 Example 8.1: Describing Diffusion Using Equilibrium Thermodynamics

For a spontaneous diffusion process between the environment and a system the diffusion process seeks to maximize

$$S_{\text{total}} = S(U,n) + S_{\text{env}}(U_{\text{env}}, n_{\text{env}}) \tag{8.1}$$

where for a combined first and second law equation we find

$$dS_{\text{total}} = \left(\frac{1}{T} - \frac{1}{T_{\text{env}}}\right) dU - \left(\frac{\mu}{T} - \frac{\mu_{\text{env}}}{T_{\text{env}}}\right) dn. \tag{8.8}$$

If the system's chemical potential is greater than the environment's chemical potential, $\mu > \mu_{env}$, then this dictates that $dn > 0$. The particles will then flow from the higher-chemical-potential region to the lower-chemical-potential region. Equilibrium occurs when the entropy is maximum $T = T_{\text{env}}$ and $\mu = \mu_{\text{env}}$, so that $dS_{\text{total}} = 0$ where diffusion ceases.

8.3 Describing Diffusion Using Probability

The probability P of finding diffusion particles a distance x from the point of initial highest concentration taken as the origin, where one can center the mean, can be expressed as a Gaussian distribution

$$C(x,t) = \frac{Q}{\sqrt{4\pi Dt}} \exp\left(-\frac{x^2}{4Dt}\right) \tag{8.14}$$

where σ is found from substitution using

$$\sigma^2 = 2Dt. \tag{8.12}$$

The result is the solution to the diffusion equation where in one dimension, the diffusion equation satisfies

$$\frac{\partial}{\partial t} C(X,t) = D(t) \frac{\partial^2}{\partial X^2} C(X,t). \tag{8.15}$$

8.4 Diffusion Acceleration Factor with and without Temperature Dependence

The diffusion coefficient itself is found to have Arrhenius temperature dependence

$$D(T)=D_0\exp\left\{-\frac{\Delta}{RT}\right\}. \tag{8.16}$$

The diffusion acceleration factor will vary according to the diffusion rate ratio

$$\mathrm{AF}_D=\frac{\text{Diffusion rate}_{\text{stress}}}{\text{Diffusion rate}_{\text{use}}}=\mathrm{AF}_T\mathrm{AF}_x \tag{8.17}$$

where AF_T is given earlier in Equation (5.21) and AF_x is the acceleration factor due to the concentration gradient, defined:

$$\mathrm{AF}_x=\frac{\left(\dfrac{\partial^2 C(x,t)}{\partial x^2}\right)_{\text{stress}}}{\left(\dfrac{\partial^2 C(x,t)}{\partial x^2}\right)_{\text{use}}}. \tag{8.18}$$

8.5 Diffusion Entropy Damage

Entropy for a Gaussian diffusion system is only a function of its variance σ^2. For a system that is diffusing over time, the entropy damage can then be measured as

$$\Delta S_{\text{damage}}=S_{t2}(X)-S_{t1}(X)=\Delta S(t_2,t_1)=\frac{1}{2}\log\left(\frac{\sigma_{t2}^2}{\sigma_{t1}^2}\right). \tag{8.21}$$

Substituting from Equation (8.12), $\sigma^2=2Dt$, we write

$$\Delta S_{\text{damage}}=\frac{1}{2}\log\left(\frac{t_2}{t_1}\right). \tag{8.22}$$

This is an important result and agrees with previous theory that was presented in the TAT model described aging in log time for thermally activated processes.

8.5.1 *Example 8.2: Package Moisture Diffusion*

Semiconductor package moisture is found experimentally as

$$\frac{\sigma^2}{2}=0.85L^2. \tag{8.23}$$

Using the time-dependent form of σ with temperature dependence, the equation has the following time-dependent power law

$$L = \sqrt{\frac{0.85}{D_0}} \exp\left(\frac{\Phi}{2RT}\right) \sqrt{t}. \tag{8.25}$$

We have noted in Section 6.2.1 (see Figure 6.3) that power laws of the form t^k, where $0 < k < 1$, can be modeled with a log(time) form (see Equations (6.9) and (6.10)). That is, through proper modeling we anticipate that a TAT model can be found for this thermally activated process in the form

$$L \approx A(T) \ln(1 + a t). \tag{8.26}$$

Reference

[1] Morse, P.M. (1969) *Thermal Physics*, Benjamin/Cummings Publishing, New York.

9

How Aging Laws Influence Parametric and Catastrophic Reliability Distributions

9.1 Physics of Failure Influence on Reliability Distributions

Reliability distributions are actually designed to fit regions of the bathtub curve shown in Figure 9.1. An overview of key reliability statistics is provided in Special Topics A to aid the reader.

For example, the Weibull distribution is basically a power law over time (see Special Topics A.4.2). If we were to invent a distribution based on wear-out for example, we might use a failure rate $\lambda(t)$ proportional to time raised to a power >1. For example, wear-out on a particular device may fit a power law with time squared as (Figure 9.2)

$$\lambda = \lambda_0 t^2. \tag{9.1}$$

This is essentially a Weibull failure rate, although the actual Weibull failure rate is written in a more sophisticated way as (see Special Topics A.4.2):

$$\lambda(t) = \frac{\beta}{\alpha^\beta}(t)^{\beta-1}. \tag{9.2}$$

This is still a power law and, in this case, β would equate to 3. Note α is the characteristic life. Many reliability engineers tend to favor the Weibull distribution because of the physical significance of the β parameter. That is, the power exponent in the distribution helps to determine the portion of the bathtub curve that we are in. If we fit our life test data to the Weibull model and $\beta = 1$, our data appear to be in the steady-state portion of the bathtub curve; $\beta > 1$ indicates wear-out; and $\beta < 1$ indicates infant mortality as shown in Figure 9.1. Often semiconductor engineers work with the lognormal distribution [1,2].

Thermodynamic Degradation Science: Physics of Failure, Accelerated Testing, Fatigue, and Reliability Applications, First Edition. Alec Feinberg.

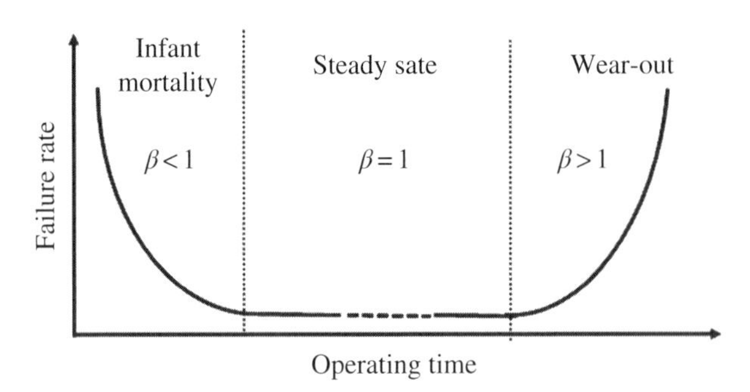

Figure 9.1 Reliability bathtub curve model

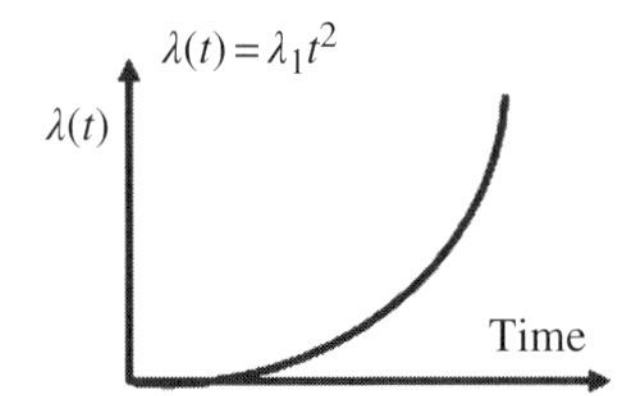

Figure 9.2 Power law fit to the wear-out portion of the bathtub curve

True statisticians will likely tell you that whichever distribution fits the data best will provide the most accurate assessment. However, we might ask a deeper question: what is the physics of failure influencing the failure rate distribution? In this chapter we will examine how the physics of the aging laws influence reliability distributions. Perhaps if we have a reason to use a particular distribution, from a physics standpoint, it might influence our decision on how we analyze our data. Aging laws have parameters. Therefore, our initial discussion will revolve around parametric reliability distributions. Once these are established we can often infer how the catastrophic distribution will follow.

9.2 Log Time Aging (or Power Aging Laws) and the Lognormal Distribution

We know from production that parameters tend to be normally distributed. For example, we can measure the strength distribution of say Young's modulus on numerous metal parts of the same type. The modulus will of course not all be exactly the same on each part, but will vary with metallurgical issues and geometric and assembly variations. That said, they are likely to be normally distributed (see Special Topics A.4.3). Now we might note that we have described physical phenomenon such as creep with the thermally activated time-dependent (TAT) model as having a log-time aging form.

We would like to illustrate that if parts are normally distributed and age in log-time, then their failure rate is lognormal (see Special Topics A.4.4). Furthermore, since we have demonstrated that power laws (such as that shown in Equation (9.3)) where the aging exponent for time K is between

0 and 1 can also be modeled as aging in log-time (see Figure 6.3), this can also apply to power laws that may be better described by log-time aging.

To that end, the general form of the TAT model is (see Equation (6.7)):

$$P = A\ln(1+bt) \approx Ct^K \quad \text{where } 0 < K < 1 \tag{9.3}$$

and to simplify, when $bt >> 1$ we can write

$$P \approx A\ln(bt). \tag{9.4}$$

Here P is the parameter of interest, such as creep, beta transistor aging, or perhaps crystal frequency drift, and so forth.

In order to have parametric failure, one needs a definition for failure. To this end it is customary to define a parametric failure threshold. That is, when a component ages, one of its key parameters drifts out of specification. This value can be used as the failure threshold. For example, transistor degradation in power loss is often taken as a loss of 10 or 20% of the original value. The figure below depicts a key parameter reaching a failure threshold of $P = 1.37$ at time $t = 3$. Here time units are not defined but are usually in hours or months and so forth (Figure 9.3).

When manufactured parts are normally distributed, a parameter of interest can be statistically assessed using Gaussian probability density function (PDF) (Special Topics A.4.3):

$$g(p,t) = \frac{1}{\sigma\sqrt{2\pi}}\exp\left[-\frac{1}{2}\left(\frac{p(t)-\bar{p}(t)}{\sigma}\right)^2\right] \tag{9.5}$$

where p is the parameter of interest. Now consider that the parameter is aging according to a log-time equation such as Equation (9.3); its time dependence must then be lognormally distributed, that is, we have from Equations (9.4) and (9.5):

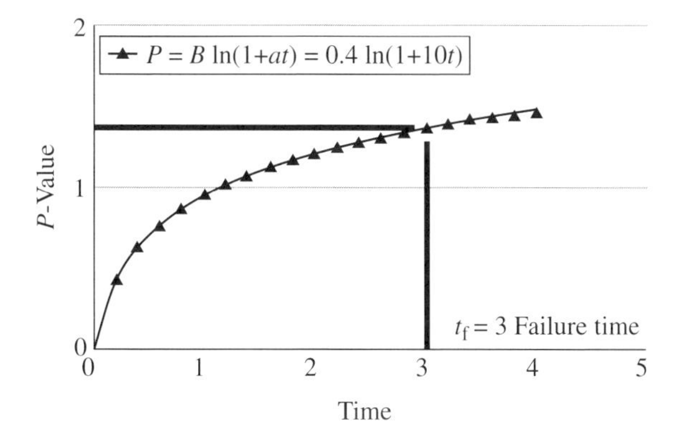

Figure 9.3 Log time aging with parametric threshold t_f

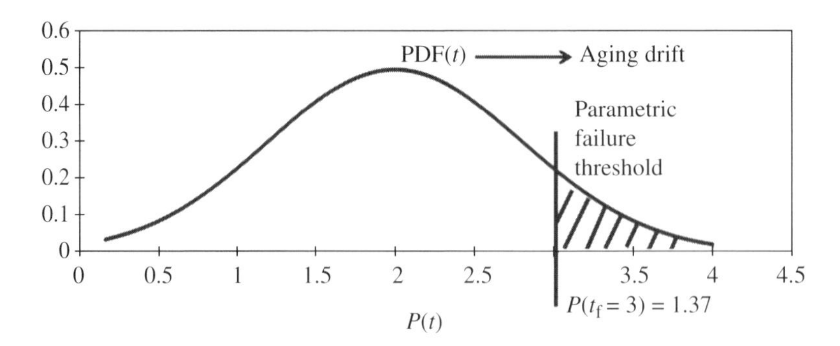

Figure 9.4 PDF failure portion that drifted past the parametric threshold

$$g(\ln t : t_{50}, \sigma) = \frac{1}{\sigma\sqrt{2\pi}} \exp\left[-\frac{1}{2}\left(\frac{\ln t - \ln t_{50}}{\sigma}\right)^2\right] \tag{9.6}$$

where for purposes of illustration in Equation (9.4) we have let $A = b = 1$ (Figure 9.4).

It is customary to change variables so that we may formally obtain the lognormal distribution for the above equation, then

$$g(\ln t) d\ln t = g(\ln t)\frac{d\ln t}{dt} dt = g(\ln t)\frac{dt}{t} = \ln g(t) dt. \tag{9.7}$$

With this change of variables we can now write

$$\ln g(t : t_{50}, \sigma) = f(t : t_{50}, \sigma) = \frac{1}{\sigma t\sqrt{2\pi}} \exp\left[-\frac{1}{2}\left(\frac{\ln t - \ln t_{50}}{\sigma}\right)^2\right] \tag{9.8}$$

where the function $f(t{:}t_{50},\sigma)$ is the lognormal PDF (see Special Topics A.4.4).

Reliability life test data is plotted using a function related to the PDF, called the cumulative distribution function (CDF) $F(t)$. They are related as

$$f(t) = \frac{dF(t)}{dt}. \tag{9.9}$$

The CDF for the lognormal distribution can be written in closed form with the help of the error function (erf) as

$$F(t) = \frac{1}{2}\left[1 + erf\left(\frac{\ln t - \ln t_{50}}{\sqrt{2}\sigma}\right)\right]. \tag{9.10}$$

Often, one writes the lognormal mean as

$$\mu = \ln(t_{50}) \tag{9.11}$$

and the dispersion is assessed graphically as

$$\sigma = \ln(t_{50}/t_{16}). \tag{9.12}$$

Thus, the physical implications can be related to log-time aging similar to a TAT model described in Chapters 6 and 7.

When a manufactured part has a key parameter that is distributed normally and ages in log-time (see Figure 9.4), its failure rate is generally lognormally distributed [2,3]. This is likely the case for power-law aging models as well as when the power $0 < k < 1$ in Equation (9.3).

Although we have described this for parametric failure, it can be argued that many catastrophic failure mechanisms dominated by log-time aging will also fall into this category. For example, if a transistor is aging most of its lifetime in log-time then suddenly fails catastrophically due to an underling log-time aging mechanism such as gate leakage (see Section 7.4.1), then the transistor's failure distribution is likely lognormal. The parametric threshold in this case resulted in a true catastrophic failure event with most of its lifetime aging logarithmically in time.

We exemplify with the TAT model, writing an aging parameter P in log-time aging form:

$$\bar{P} = \bar{A}\langle \ln(1+bt)\rangle_{ave} \cong \bar{c} + \bar{A}\ln(t_{50}) \tag{9.13}$$

where the approximation is for $bt >> 1$; $\bar{c}, \bar{A}$ are average values; and $\bar{c} = \bar{A}\langle \ln b\rangle_{\text{ave}}$ and $\ln(t_{50})$ is the mean of $[\ln(t)]$ failure time. Then Equation (9.8) for the parametric PDF becomes

$$f(t:t_{50},\sigma) = \frac{1}{\sigma t\sqrt{2\pi}} \exp\left(-\frac{1}{2}\left\{\frac{[c + A\ln(t)] - (\bar{c} + \bar{A}\ln(t_{50}))}{\sigma}\right\}^2\right). \tag{9.14}$$

The CDF is then

$$F(t) = \frac{1}{2}\left(1 + \text{erf}\left\{\frac{[c + A\ln(t)] - [\bar{c} + \bar{A}\ln(t_{50})]}{\sqrt{2}\sigma}\right\}\right). \tag{9.15}$$

9.3 Aging Power Laws and the Weibull Distribution: Influence on Beta

Most parametric aging laws have a power-law dependence. Consider creep as an example:

$$\Delta\varepsilon = at^n \tag{9.16}$$

where $\Delta\varepsilon$ is the creep strain; t is time; and a and n are constants of the creep model [4]. This simple equation can actually model both the primary and secondary creep phases, as well as roughly the third-stage tertiary creep phase as shown in Figure 9.5.

Now we would like to provide some new understanding of the Weibull distribution, how underlying aging laws might influence the distribution, or how analysis might help us in

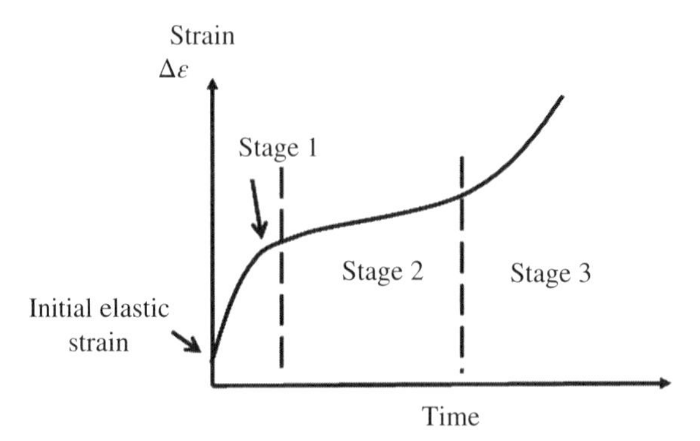

Figure 9.5 Creep curve with all three stages

determining an aging law (not found in other books). As a point of reference, we write the popular Weibull failure rate as (Special Topics A.4.2)

$$\lambda(t) = \frac{\beta}{\alpha^{\beta}}(t)^{\beta-1}. \tag{9.17}$$

For the traditional Weibull model, $\beta < 1$ is infant mortality; $\beta = 1$ is steady state; and $\beta > 1$ is wear-out.

There are traditional functions to help obtain the failure rate in reliability statistics. The functional definition for the instantaneous failure rate is defined with

$$\lambda(t) = -\frac{d\ln R(t)}{dt} = \frac{f(t)}{R(t)} = \frac{f(t)}{1-F(t)} \tag{9.18}$$

where $R(t)$ is the reliability function, $f(t)$ is the PDF, and $F(t)$ is the CDF.

However, for what we wish to do we are going to start off with a simplified definition for the average expected failure rate as:

$$\bar{\lambda}(t) = \frac{\Delta E}{\Delta t} \tag{9.19}$$

where ΔE is the expected fractional units that fail in the time interval Δt. Then in the limit as Δ becomes infinitesimally small, we write the failure rate as

$$\lambda(t) = \lim_{\Delta \to 0} \frac{\Delta E}{\Delta t} = \frac{dE}{dt}. \tag{9.20}$$

Let's now look at an oversimplified parametric aging power law form for the three stages of creep (see Equation (4.13) for a more detail model)

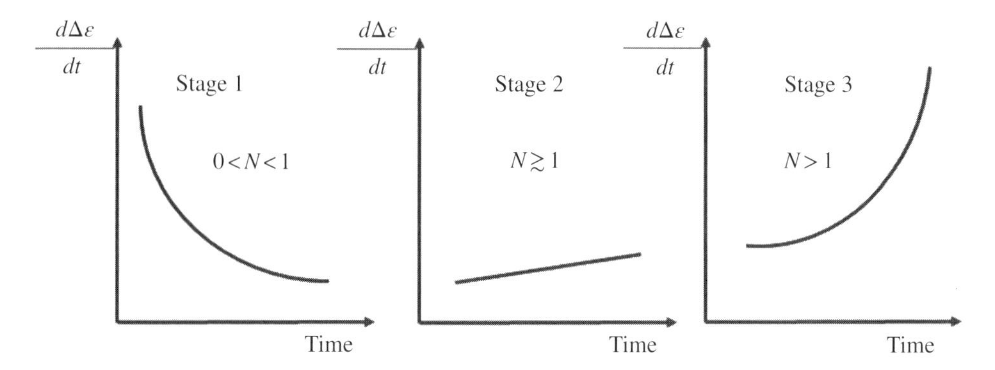

Figure 9.6 Creep rate power law model for each creep stage, similar to the bathtub curve in Figure 9.1

$$\Delta\varepsilon = \varepsilon_0 t^N \begin{cases} N < 1 & \text{stage 1} \\ N = 1 & \text{stage 2} \\ N > 1 & \text{stage 3} \end{cases} \tag{9.21}$$

There are numerous time-dependent creep models that are commonly used that have more complex forms and are better suited to model creep. For example, in Equation (4.13) have seen that $\varepsilon_0 = A\sigma^b$ indicating different stresses affect the creep slope (see Figure 9.6). This particular model is oversimplified, but it is roughly capable of modeling all three stages of creep shown in Figure 9.5. This oversimplified power-law form is very instructional as there are numerous similar power aging laws of this type in physics of failure applications. The three stages of creep are shown in Figure 9.5.

When $0 < N < 1$ it models Primary Stage 1; when $N = 1$ it models Secondary Stage 2; and when $N > 1$ it models Tertiary Stage 3. The creep rate is defined

$$\frac{d\Delta\varepsilon}{dt} = \varepsilon_0 N t^{N-1}. \tag{9.22}$$

Now using the different power-law values for creep, we can plot the creep rate curve as shown in Figure 9.6. Interestingly enough, the bathtub curve in Figure 9.1 has a similar shape to the creep rate curve shown in Figure 9.6. Note that Stage 2, although not flat like the idealized bathtub curve, there are some situations reported that this steady-state area of the bathtub curve has a small increasing failure rate.

What we are therefore tempted to do is try and merge the physics of creep to the statistical failure rate equations and make inferences. Although it is seemingly discomfited, we will see that it is insightful.

Let us start by saying that for any creep phase, we can have a parametric failure corresponding to the parametric failure threshold t_{failure} so that the time to failure in general will be given by

$$\Delta\varepsilon_{failure}(t) = \varepsilon_0 t_{\text{failure}}^N \tag{9.23}$$

where we assume the following parametric treatment. For example, in testing when a device passes the failure threshold $\Delta\varepsilon_{\text{failure}}$, corresponding to a time $\text{t}_{\text{failure}}$, we count the device as a failure and proceed to perform some sort of traditional reliability catastrophic type of analysis to find its failure distribution and failure rate based on the time to failure for each device that passes the creep threshold. We have no idea of the life test parametric distribution, however. If it were normal for example, we would have a mean time to failure and a spread in the times to failure given by sigma.

However, in this discussion we would like to proceed and make inferences from the aging law on how it influences the statistics. The expected fraction of devices that will fail ΔE ($\Delta\varepsilon_{\text{failure}}(t)$) in the time interval Δt then must be a function of the aging law so that the failure rate as we have defined it above is

$$\lambda(t) = \frac{dE}{d(\Delta\varepsilon_{\text{failure}}(t))}\frac{d(\Delta\varepsilon_{\text{failure}}(t))}{dt} = g(E)\varepsilon_0 N t_{\text{failure}}^{N-1} \tag{9.24}$$

where we have let

$$g(E) = \frac{dE}{d(\Delta\varepsilon_{\text{failure}}(t))}. \tag{9.25}$$

If we assume a Weibull distribution for the parametric failure rate, we can make some observations. By direct comparison to the traditional Weibull parameters between Equations (9.17) and (9.24), we conclude

$$N = \beta \text{ and } g(E)\varepsilon_0 = (1/\alpha)^{\beta}. \tag{9.26}$$

So in this model if $0 < N < 1$, say $\beta = N = 1/2$, indicating that creep is in the Primary Creep Stage 1, then we are also in the infant mortality region. This is reasonable as it indicates early failure. If $N = 1$, then we have a constant creep rate which is in the Secondary Creep Stage 2. This is also associated with the steady-state region of the bathtub curve as $\beta = 1$. Finally, if $N = \beta > 1$, we reach the Tertiary Creep Stage 3 and are in the wear-out phase of the bathtub curve. Therefore, the physics for creep rate matches the statistics reasonably well.

Essentially we have made direct comparisons between the creep rate in Figure 9.6 and the failure rate in Figure 9.1, finding that $N \sim \beta$. It is therefore likely that for numerous aging power laws, when carefully modeled, the power exponent, can be directly tied to the value of the Weibull β. We have now connected both the Weibull model and the lognormal model for physical aging laws.

In catastrophic analysis, it is customary to assign α to a value of the aging equation, for example

$$\alpha = t_{\text{final}} = \left(\frac{\Delta\varepsilon_{\text{failure}}}{\varepsilon_0}\right)^{1/N}. \tag{9.27}$$

This is a number so we can just keep it in mind. It is evaluated at the failure time for parametric failure.

This now begs the question, when we do a Weibull catastrophic failure analysis (i.e., hard failure as compared to soft parametric failure) and find the parameters α and β, is there an underlying aging law of the form in Equation (9.23) that we can associate with the Weibull parameters?

It is likely that there is not a perfect answer for this question. First it would require a simplified aging law similar to Equation (9.23). Then it requires that the dominant part of a product's lifetime was due to parametric well-behaved failure that can be modeled by the aging power law. Lastly, the catastrophic failure event would be fairly abrupt so that the aging law is a good approximation of the lifetime. That being said, it is still best to do a multivariable Weibull analysis so the parameters N, β, and α can be found from the failure analysis.

9.4 Stress and Life Distributions

It is instructive to illustrate how to incorporate a stress model into a life distribution. We have done this already from first principals. However, we would like to illustrate it for commonly used time-to-failure stress equations. We will illustrate this for both the power law form and the Arrhenius function and how these are incorporated into the CDF and PDF for the lognormal distribution (Figure 9.7).

Consider the vibration stress causing creep or fatigue ($\sigma(W)$) and the Arrhenius temperature time to failure model given earlier in Equations (4.46) and (5.22), respectively. The time to failure is written in linear form as

$$\ln(t_{\text{failure}}) = C + \frac{E_{\text{a}}}{k_{\text{B}}T}. \tag{9.28}$$

Then for a vibration stress we would consider Equations (4.38) and (4.46) in linear form

$$\ln(t_{\text{failure}}) = C = M_b \ln(W). \tag{9.29}$$

where $M_b = -(b/2)\ln(\beta/f)$ (Equation (4.46)).

Experimentally, the time to failure can be assessed at any time. For the lognormal distribution, these parameters apply to the median time to failure, $t_{\text{failure}} = t_{50}$. This allows for a direct substitution into the lognormal distribution functions of Figures A.8 and A.9. Inserting the Arrhenius function into the PDF gives us

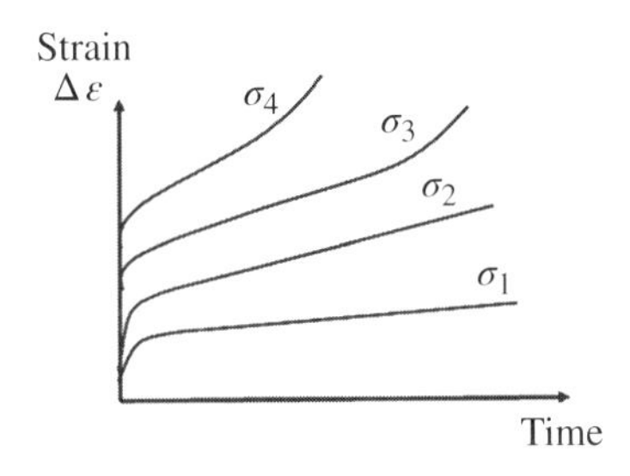

Figure 9.7 Creep strain over time for different stresses where $\sigma_4 > \sigma_3 > \sigma_2 > \sigma_1$

$$f(t,T)=\frac{1}{\sigma_t(T)t\sqrt{2\pi}}\exp\left\{-\frac{1}{2}\left[\frac{\ln(t)-\left(C+\dfrac{E_a}{k_B T}\right)}{\sigma_t(T)}\right]^2\right\}. \tag{9.30}$$

For the vibration model, this is

$$f(t,W)=\frac{1}{\sigma_t(W)t\sqrt{2\pi}}\exp\left\{-\frac{1}{2}\left[\frac{\ln(t)-(C-M_b\ln(W))}{\sigma_t(W)}\right]^2\right\}. \tag{9.31}$$

Similarly, inserting the Arrhenius model into the CDF (using the erf form in Equation (A.23)) yields

$$F(t,T)=\frac{1}{2}\left\{1+\operatorname{erf}\left[\frac{\ln(t)-\left(C+\dfrac{E_a}{k_B T}\right)}{\sqrt{2}\sigma_t(T)}\right]\right\}. \tag{9.32}$$

For the vibration model, this is

$$F(t,W)=\frac{1}{2}\left\{1+\operatorname{erf}\left[\frac{\ln(t)-(C-M_b\ln(W))}{\sqrt{2}\sigma_t(W)}\right]\right\}. \tag{9.33}$$

Similar expressions can be found for the CDF and PDF of any life distribution function when t_{failure} is appropriately found [2]. As an exercise, the reader might wish to apply this methodology for the Weibull CDF and PDF. (Hint: assume that $t_{0.632}=t_{\text{failure}}$; this is the characteristic life α_w in the table.)

9.4.1 *Example 9.1: Cumulative Distribution Function as a Function of Stress*

For the vibration function, let $C=-7.82$, $M_b=4$, and find $F(t,W)$ for $t=10$ years and $W=0.0082\ \text{G}^2/\text{Hz}$. Find F at 10 years. Use $\sigma=2.2$ for your estimate. If the stress level is reduced by a factor of 2, what is F?

Solution

Inserting these values into the CDF above gives us

$$F(t,W)=\frac{1}{2}\left\{1+\operatorname{erf}\left[\frac{\ln(87600)-(-7.82-4\ln(0.0082))}{\sqrt{2}(2.2)}\right]\right\}$$

or

$$F(87600,0.0082)=\frac{1}{2}\left[1+\operatorname{erf}\left(\frac{-0.0139}{\sqrt{2}(2.2)}\right)\right]=\frac{1}{2}\left[1-\operatorname{erf}\left(\frac{0.0139}{\sqrt{2}(2.2)}\right)\right]=0.497.$$

Thus, at this stress level, 49.7% of the distribution is anticipated to have failed in 10 years. (Note that in the above derivation the erf values can be found from tables or in Microsoft Excel type "erf(0.00447)" to obtain the above value.) If the stress level is reduced by a factor of 2, then $W = 0.0041$ G^2/Hz. The anticipated percent failure at 10 years is reduced to $F(87\,600, 0.0041) = 10.27\%$.

9.5 Time- (or Stress-) Dependent Standard Deviation

Most models assume that β for the Weibull or σ for the normal or lognormal distributions are independent of time and/or stress. This may not always be the case, especially in parametric analysis. Figure 9.8 illustrates parametric crystal frequency drift of a distribution of oscillators tested at 120°C [3]. This type of plot is a cumulative probability plot versus frequency. Each fitted line shows the observed mean and standard deviation observed at 400, 600, and 1000 h. The slope of each time group leads to the standard deviation; as can be seen, it varies greatly.

Such parametric dependence leads to a time dependence on sigma and likely also in the catastrophic case for crystal oscillator frequency drift. Usually, there is not enough data to determine this dependence and σ is treated as a constant over time and/or stress.

The simplest way to model a time or stress-dependent standard deviation from the data such as those shown in Figure 9.8 is to first determine the standard deviation for each distribution.

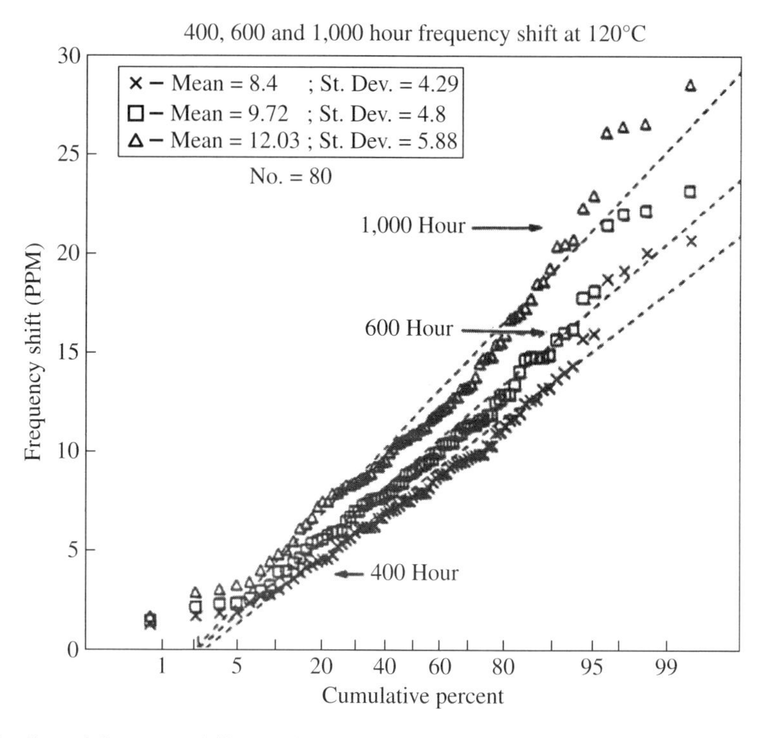

Figure 9.8 Crystal frequency drift showing time-dependent standard deviation. Source: Feinberg [3]. Reproduced with permission of IEEE

This is usually given in the software but can also be easily read off the graph, for example at 400 h we have

$$\sigma_{400\text{h}} = f_{50\%} - f_{16\%}. \tag{9.34}$$

Once we have the standard deviation for each distribution over time, that is, $\sigma(400)$, $\sigma(600)$, $\sigma(1000)$ at times $t = 400$, $t = 600$, and $t = 1000$ h, we can then plot σ versus time on a graph and establish the relationship by fitting this data. For example, if it is linear then a likely model would lead to a straight line fit for these points as [2]

$$\sigma(t) = A + Bt. \tag{9.35}$$

If σ is stress dependent, we can follow the same procedure. We can also do this for a Weibull distribution if the data show, for example, that β is stress or time dependent.

Summary

9.1 Physics of Failure Influence on Reliability Distributions

Reliability distributions are actually designed to fit regions of the bathtub curve. It was noted that the Weibull distribution is basically a power law over time.

$$\lambda(t) = \frac{\beta}{\alpha^{\beta}}(t)^{\beta-1}. \tag{9.2}$$

This can fit any region of the bathtub curve. When $\beta = 1$, our data appear to be in the steady-state portion of the bathtub curve; $\beta > 1$ indicates wear-out; and $\beta < 1$ indicates infant mortality as shown in Figure 9.1.

9.2 Log Time Aging and the Lognormal Distribution

In this section we illustrated that if parts are normally distributed and age in log-time, then their failure rate is lognormal. We started with a simplified TAT model with $bt >> 1$ written as

$$P \approx A \log(bt) \tag{9.4}$$

where P is the parameter of interest such as creep, beta transistor aging, or crystal frequency drift and so forth. When manufactured parts are normally distributed, a parameter of interest can be statistically assessed using Gaussian PDF $g(p, t)$, defined:

$$g(p,t) = \frac{1}{\sigma\sqrt{2\pi}} \exp\left[-\frac{1}{2}\left(\frac{p(t)-\bar{p}(t)}{\sigma}\right)^2\right]. \tag{9.5}$$

By inserting the simplified TAT model and with a change of variables we were able to obtain the lognormal PDF distribution functional form

$$\ln g(t:t_{50},\sigma)=f(t:t_{50},\sigma)=\frac{1}{\sigma t\sqrt{2\pi}}\exp\left[-\frac{1}{2}\left(\frac{\ln t-\ln t_{50}}{\sigma}\right)^2\right]. \quad (9.8)$$

9.3 Aging Power Laws and the Weibull Distribution: Influence on Beta

We studied an oversimplified parametric aging power law for the three stages of creep

$$\Delta\varepsilon=\varepsilon_0 t^N \begin{cases} N<1 & \text{stage 1} \\ N=1 & \text{stage 2} \\ N>1 & \text{stage 3} \end{cases} \quad (9.21)$$

It was noted that the bathtub curve in Figure 9.1 has a similar shape to the creep rate curve. This led to an investigation of the creep failure rate:

$$\lambda(t)=\frac{dE}{d(\Delta\varepsilon_{\text{failure}}(t))}\frac{d(\Delta\varepsilon_{\text{failure}}(t))}{dt}=g(E)\varepsilon_0 N t_{\text{failure}}^{N-1}. \quad (9.24)$$

By direct comparison to the traditional Weibull parameters between Equations (9.17) and (9.24), we concluded

$$N=\beta \text{ and } g(E)\varepsilon_0=(1/\alpha)^\beta. \quad (9.26)$$

In this model if $0<N<1$, say $\beta=N=1/2$, creep is in the Primary Creep Stage 1 and we are also in the infant mortality region. This is reasonable as it indicates early failure. If $N=1$, then we have a constant creep rate which is in the Secondary Creep Stage 2. This is also associated with the steady-state region of the bathtub curve as $\beta=1$ bathtub curve. The physics and the statistics therefore match reasonably well.

Essentially we have made direct comparisons between the creep rate in Figure 9.6 and the failure rate in Figure 9.1, finding that $N\sim\beta$.

This begs the question, when we do a Weibull catastrophic failure analysis (i.e., hard failure as compared to soft parametric failure) and find the parameters α and β, is there an underlying aging law that we can associate with the Weibull parameters?

It is likely that there is not a perfect answer for this question. However, some insight was suggested.

We have now connected both the Weibull model and the lognormal model to physical aging laws.

9.5 Time- (or Stress-) Dependent Standard Deviation

Most models assume that β for the Weibull or σ for the normal or lognormal distributions are independent of time and/or stress. This may not always be the case, especially in parametric analysis.

For example, we showed graphically how one could model σ; if it is linear, then a likely model would lead to a straight line fit for these points as [2]

$$\sigma(t) = A + Bt. \tag{9.35}$$

References

[1] O'Connor, P. and Kleyner, A. (2012) *Practical Reliability Engineering*, 5th edn, John Wiley & Sons Ltd, London.
[2] Feinberg, A. and Crow, D. (eds) *Design for Reliability*. M/A-COM 2000. CRC Press, Boca Raton, 2001.
[3] Feinberg, A. (1992) Gaussian parametric failure rate model with applications to quartz-crystal device aging. *IEEE Transaction on Reliability*, **41**, 565.
[4] Collins, J.A., Busby, H. and Staab, G. (2010) *Mechanical Design of Machine Elements and Machines*, 2nd edn, John Wiley & Sons, Inc., New York.

10

The Theory of Organization: Final Thoughts

This chapter is a late entry, where we present final thoughts to help serve as a two page summary for the notions in this book. To do so we present the Theory of Organization for stability, reliability, and quality. Since we are often interested in the solution to the issues of degradation, the best general top-down approach that we can offer is to assert the following.

> *The Theory of Organization for stability, reliability, and quality: For a system, subsystem, component, material, or process, in general, the higher the organization, the more likely is the probability of success over time.*

In fact, statistically, reliability is the probability of success over time (see Equations (A.2) and (A.10)). If $P_i = 1/N$ for N different microstates ($i = 1, 2, \ldots, N$) then, as we have stated in Equation (2.31), our measure of disorder for entropy is

$$S = -k_B \sum_i \left(\frac{1}{N}\right) \log\left(\frac{1}{N}\right) = k_B \ln N. \tag{10.1}$$

This helps our notion of entropy, disorder, and, of course, order over time. The larger the number of N_i microstates that can be occupied, the greater is the potential for disorder. We therefore seek to simplify the number of possible states in a system and the capability of the system to fill the microstates. For example, we might model the capability of a system to fill the microstates in time and cause entropy damage as:

$$N(t) = N_{\mathrm{Max}}(1 - \exp(-\tau t)). \tag{10.2}$$

Thermodynamic Degradation Science: Physics of Failure, Accelerated Testing, Fatigue, and Reliability Applications, First Edition. Alec Feinberg.

In this model, there are two variables to control disorder: N_{max} and the time constant τ. The variable N_{max} applies to our everyday notions in design and manufacturing. Reducing the potential for disorder through organization is equivalent to designing for a smaller N_{max}. Simplicity, reducing complexity, and structural order can be key for the probability of success for our systems, subsystems, components, materials, and processes. This does not necessarily mean we cannot have complexity. However, we would then need to focus on designing for small time constants τ. As we have seen, one should seek to maximize the free energy of a system and minimize variability in a system or a process. Processes often relate to the normal distribution (see Equation (2.35)); variations in our design and defects in our materials increase the potential for disorder, creating opportunities for entropy damage. Variability is therefore a strong measure of disorder. As we have noted in Chapter 2, variability can promote noise. We know that materials such as crystals and highly repetitive structures such as metals and diamonds (polynomial structures) are often the strongest and most reliable materials. Their redundant structure and designs can minimize the potential for entropy damage for materials and structures. Their free energy, that is, capacity to perform useful work, is higher than for amorphous materials or untidy structures. There are of course exceptions. We seek to dampen resonance in circuit boards to reduce issues of vibration failure. In many cases, common sense prevails: poor design creates loss of order and is strongly related to the probability of degradation occurring over time. Every chapter in this book supports the theory of organization in some way to increase our probability of success in preventing degradation, and is a validation of this theory to aid in our understanding of how to improve stability, quality, and reliability for our products.

Special Topics A

Key Reliability Statistics

A.1 Introduction

Reliability statistical analysis can be complex. However, the basic statistics needed for industry are not too difficult and we will provide an overview of the key statistics needed for the reader. This will help in understanding the full scope of the science of degradation and its analysis. We will overview a number of methods that are common throughout the industry.

It can be somewhat incomplete to understand the physics of degradation and not understand reliability. This is because understanding and preventing failures from occurring is an inexact science. This means we have to deal with distributions and probability of failure occurring. It is not enough to understand the failure mechanism. We need to ask the following questions.

- What is the failure mechanism?
- What is the probability of it occurring?
- Can we demonstrate with a certain probability that a high failure rate will not occur on a product?
- How can we verify that failure will not occur if the product is to be in the field for 10 years?
- What types of accelerated test do we do?
- What is the sample size that is required?
- What is the confidence in our estimates?

A.1.1 Reliability and Accelerated Testing Software to Aid the Reader

It is helpful to use software in understanding reliability and accelerated testing analysis in Special Topics A and B. Free reliability statistics and accelerated testing software is available

Thermodynamic Degradation Science: Physics of Failure, Accelerated Testing, Fatigue, and Reliability Applications, First Edition. Alec Feinberg.

to aid the reader at the author's website, www.dfrsoft.com, to try for all the examples in Special Topics A and B [1]. This chapter provides reference to the module and row number that the reader can go to in the software to follow the examples. For example: (dfrsoft, module 2, row 13). The software is in friendly Excel format with reference modules (see menu) that includes help with: reliability conversions; reliability plotting (Weibull, lognormal, ...); system reliability; distributions; field returns; acceleration factors; test plans; environmental profiling; reliability growth; reliability predictions; parametric reliability; availability and sparing; derating; physics of failure; process capability index (Cpk) analysis; lot sampling; statistical process control (SPC); thermal analysis; shock and vibration; corrosion; and design of experiment.

A.2 The Key Reliability Functions

There are a number of key reliability functions that help define reliability. The key reliability functions that we will be referring to in this chapter are listed below.

- *Cumulative Distribution Function (CDF) $F(t)$*: This is an important function for data analysis as we will show in this chapter. Experimentally, it is the cumulative percent failure at each observed failure time when plotted versus time. It therefore gives a probability of a component failing at time t. We will provide examples of how to find this in Sections A.7.2 (Example A.6) and A.7.3 (Example A.7). In terms of the reliability $R(t)$, we write

$$F(t) = 1 - R(t). \tag{A.1}$$

- *Reliability Function $R(t)$*: This is the number of units surviving at time t divided by initial number of units. Another way of saying this is the probability of the component surviving to time t.

$$R(t) = 1 - F(t). \tag{A.2}$$

- *Probability Density Function (PDF) $f(t)$*: Experimentally, it is the instantaneous slope at time t found on the CDF plot. One might recall how the normal distribution PDF looks like a bell shaped curve.

$$f(t) = \frac{dF(t)}{dt}. \tag{A.3}$$

- *Failure Rate $\lambda(t)$*: Often called the hazard rate or instantaneous failure rate, this is expressed in terms of $f(t)$ and $R(t)$ as (see also average failure rate):

$$\lambda(t) = \frac{f(t)}{R(t)} = -\frac{1}{R(t)}\frac{dR(t)}{dt}. \tag{A.4}$$

- *Cumulative Failure Rate $\lambda(t)_{\mathrm{cum}}$*: This is not used often in reliability. Once $F(t)$ is known however, it is simply given by

$$\lambda_{\mathrm{cum}}(t) = \frac{dF(t)}{t}. \tag{A.5}$$

- *Average Failure Rate $\lambda(t)_{\mathrm{ave}}$*: This is the fractional failures occurring or the expected fractional failures occurring in time Δt. As $\Delta \rightarrow 0$, this is equal to the hazard rate.

$$\lambda_{\mathrm{ave}}(t) = \frac{\Delta fr}{\Delta t} = \frac{\text{No. Failures/Total Number}}{\Delta t}. \tag{A.6}$$

- *Mean Time To Failure (MTTF) or Mean Time Between Failures (MTBF)*: This is the mean time that we anticipate a system will be operational. When a system is repairable we use MTBF. When it is not repairable, such as a semiconductor, we use MTTF. Note that people often use MTBF for a system or part that is not repairable! While this is of course not accurate, the reader should be aware of such common misuse of the term. When the failure rate (FR) is constant we can write

$$\text{MTBF or MTTF} = \frac{1}{\text{constant FR}} = \frac{1}{\lambda}. \tag{A.7}$$

- *Availability*: This is a term used for how available a system is in its steady-state operation mode. In simple terms, people often use what is called the "inherent availability" that is equal to the "up time"; the MTBF; and "down time," which is usually the mean time to repair (MTTR) a system. Often availability is a better number to advertise. Inherent availability is then

$$A = \frac{\text{MTBF}}{\text{MTBF} + \text{MTTR}} = \frac{\text{Up Time}}{\text{Up Time} + \text{Down Time}}. \tag{A.8}$$

In addition to these functional definitions, two key areas of reliability should be defined. These are *system* and *component* reliability. Component reliability concerns discrete items such as resistors, capacitors, diodes, etc. System reliability concerns issues of multiple discretes that make up a unit, such as hybrids, subassemblies, and assemblies. In system reliability, the whole is usually equal to the sum of the parts in terms of the failure rate, unless the system has what is called redundancy.

As we have noted in our definitions, we can talk about system reliability which is usually made up of many parts such as hybrids or assemblies, or we can talk about component reliability (resistor, integrated circuits or ICs, etc.). Each component has associated with it a failure rate that is based on analysis or historical information from prior experiments. There are a number of prediction methods that use libraries of failure rate values for different components. Two popular methods for predicting the failure rate of an assembly are Telcordia and MIL STD 217 (see dfrsoft, modules 11, 12, 13, 14). There are also a number alternative methods used today. The key for accurate predictions however is of course the library and knowledge of the stress applied to each part and the stress modeled used to assess the component failure rate. Often, these libraries and stress models are not very accurate. As we mentioned, reliability is not an exact science. It is always better to have field data to obtain a more accurate MTBF. The actual failure rate for the assembly when there is no redundancy simply sums the modeled failure rates of each component. Therefore, the failure rate for an assembly (with no redundancy) with N critical parts can be written after each component failure rate is modeled with a stress assessment as:

$$\lambda(t)_{\text{assembly}} = \sum_{i=1}^{N} \lambda_i. \tag{A.9}$$

We will not be focusing on assembly reliability. Since assemblies are made up of parts, it is important to understand component reliability first. Once one has a grasp of component reliability, there are numerous references for understanding assembly reliability [2, 3].

A.3 More Information on the Failure Rate

Failure rates can be classified into two categories:

- time-dependent failure rate $\lambda(t)$; and
- time-independent failure rate $\lambda(t) = \lambda$.

For a time-dependent failure rate, we need to specify the time at which the failure rate is given. The time-dependent failure is typically more complicated to asses. This is obtained once we have knowledge of the PDF and/or the reliability time-dependent function through the derivative as we defined in Section A.2. Experimentally, we can also obtain this from the CDF as will be exemplified in this chapter. However, the time-independent failure rate is much easier to work with so we will start with this. We can use the definitions at Equations (A.6) and (A.7) to provide a simple example (see dfrsoft, module 1, row 13 [1]) for a constant failure rate. For example, using Equation (A.7) and assuming an MTTF of 10 h, then $\lambda = 1/\text{MTTF} = 1/(10\,\text{h}) = 0.1$ fractional failure/h.

A very popular reliability metric that we will use in this chapter is failures per million hours (FMH) and failures per billion hours (FITs). To convert the above failure rate to FMH and FITs (see dfrsoft, module 1, row 13 [1]):

To convert to FMH, 0.1 fractional failure $(1 \times 10^6)/(1 \times 10^6)$ h; 100 000 FMH where the denominator $1/(1 \times 10^6)$ = FMH.
To convert to FITs, 0.1 fractional failure $(1 \times 10^9)/(1 \times 10^9)$ h; 100 000 000 FITs where the denominator $1/(1 \times 10^9)$ = FITs.

Another example on using the constant failure rate, for 2% failure (0.02 fractional failure) in 8760 h (1 year):

$$\lambda = 0.02\,\text{fraction fail}/8760\,\text{h} = 0.00000228\,\text{fraction failure per hour}.$$

Convert to failures per million:

$$\lambda = 2.28\,\text{FMH}.$$

Table A.1 Constant failure rate conversion table

FMH (fail/10^6 h)	MTBF (h)	1 year (PPM)	1 year (% failure)	10 year (PPM)	10 year (% failure)
0.001	1.00×10^9	9	0.0009	88	0.009
0.005	2.00×10^8	44	0.0044	438	0.044
0.025	40 000 000	219	0.022	2188	0.22
0.100	10 000 000	876	0.09	8722	0.87
0.260	5 000 000	1750	0.18	17 367	1.74
0.290	2 500 000	3498	0.35	34 433	3.44
1.00	1 000 000	8722	0.87	83 873	8.39
2.00	500 000	17 367	1.74	160 711	16.07
4.00	250 000	34 433	3.44	295 594	29.6
10.00	100 000	83 873	8.39	583 555	58.4

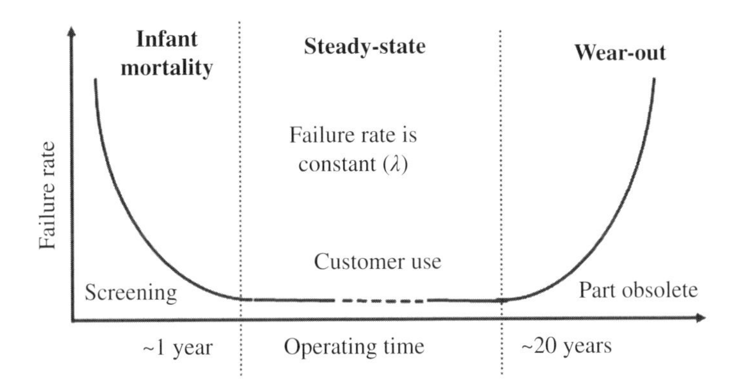

Figure A.1 Reliability bathtub curve model

Convert to MTTF = 1/failure rate = 438 000 h.

Table A.1 provides a list of some important constant failure rate metrics. The reader may wish to verify some of the numbers as an exercise (see dfrsoft, module 1, row 13 [1]).

A.4 The Bathtub Curve and Reliability Distributions

The reader is likely familiar with the bathtub curve and its regions. Many readers may not however realize that the key reliability models are closely related to this curve. It is therefore a good place to start and explain reliability models, how they are derived, and how they are used. The common bathtub curve appearing in most reliability books is shown in Figure A.1. The curve is modeled after the human mortality rate. The important thing to note is that *the key reliability failure-rate models will fit the bathtub curve or portions of it.* The main regions of the bathtub cure are as follows:

- The infant-mortality region indicates the portion of shipped product that fails in early life. This typically goes up to 1 year. This is an important region for commercial products as it is the worst possible situation to allow customers to receive products that can fail in early life. If a proper qualification is not done on the product, and potential infant mortality still exists, then companies have to come up with a screen to weed out bad products.

- The next important region is the steady-state period. This is the area where customers are expected to use the product. The failure rate is constant and it is desired that this is as low as possible.
- The wear-out region represent end-of-life failures. It depends on the product type. Passive components can last 25 years; power transistors may only last for 10 years. Often, the wear-out region is not reached when the item is out of date. Cell phones can be a common case.

There are four key reliability distributions that are commonly used: exponential; Weibull; lognormal; and normal. The former three can model all or portions of the bathtub curve, and are commonly used in catastrophic life data analysis. While the normal distribution is rarely used by reliability engineers in this way; it is well used in other areas (see Section A.6.1). Below is a summary of each distribution and some of the rationale behind it.

A.4.1 Exponential Distribution

The exponential distribution is the simplest to understand as it is a one-parameter distribution; that parameter is the failure rate which is independent, yielding a constant failure rate. It therefore models the flat part of the bathtub curve where $\lambda(t) = \lambda$ and is given by:

- Reliability function:

$$R(t) = e^{-\lambda t}. \tag{A.10}$$

- CDF (see Equation (A.2)):

$$F(t) = 1 - R(t) = 1 - e^{-\lambda t}. \tag{A.11}$$

- PDF (see Equation (A.3)):

$$f(t) = \frac{dF(t)}{dt} = \lambda e^{-\lambda t}. \tag{A.12}$$

- Failure rate (see Equation (A.4)):

$$\lambda(t) = \lambda. \tag{A.13}$$

A.4.1.1 Example A.1: Some Basic Math of the Exponential Distribution

From Equation (A.4) we can verify that $R(t)$ above satisfies the requirements for the exponential distribution, that is

$$\text{Failure rate} = \text{Constant} = \lambda = -\frac{1}{R(t)}\frac{dR(t)}{dt}.$$

Insert Equation (A.10) to see that it satisfies the result of Equation (A.13).

Another important result of the exponential distribution is noted by doing a Taylor series expansion on Equation (A.11). That is

$$F(t) = 1 - e^{-\lambda t} = 1 - (1 - \lambda t + \text{smaller order terms}\ldots) \approx \lambda t.$$

This is important to note as it is intuitive that, for a constant failure rate which has units of fractional failure per unit time, when we multiply the failure rate by time we get the fractional failure. The above approximation works when $\lambda t << 1$ (where higher-order terms are small).

A.4.1.2 Example A.2: Estimating the Number of Failures and Availability with Exponential Reliability Function

Consider a complex repairable item with a constant failure rate of 25 FMH. What is its MTBF? If 1000 units are shipped per year at a 90% duty cycle, what is the percentage that is anticipated to fail in the first year? If the MTTR is 72 h, what is the inherent availability?

The constant failure rate is then (see dfrsoft, module 1, row 13 [1]):

$$\lambda = 25\,\text{FMH}\,\frac{10^{-6}/\text{h}}{\text{FMH}} = \frac{25}{10^{6}\,\text{h}} = \frac{2.5 \times 10^{-5}}{\text{h}}.$$

With 50% duty cycle per year, the operating time is 4380 h. Then the number of failures per year is (see dfrsoft, module 1, row 13 [1]):

$$F(t) = \left[1 - \exp\left(-\frac{2.5 \times 10^{-5}}{\text{h}} 4380\ \text{h}\right)\right] \times 1000 = 104.$$

We note that $\lambda t \times 1000 = 110$, which illustrates that the approximation $F(t) \approx \lambda t$ is reasonable but not as accurate as one might like. The MTBF is

$$\text{MTBF} = \frac{1}{\lambda} = 40\,000\ \text{h}.$$

The inherent availability (see Equation (A.8)) is

$$A = \frac{\text{MTBF}}{\text{MTBF} + \text{MTTR}} = \frac{40000}{40000 + 72} = 0.998.$$

Note that this item has an MTBF that may not be competitive. However, because we can repair it in just 3 days, we might wish to only advertise the availability of 99.8%. This may be a very competitive number to tell customers, who are likely only interested in the operational availability.

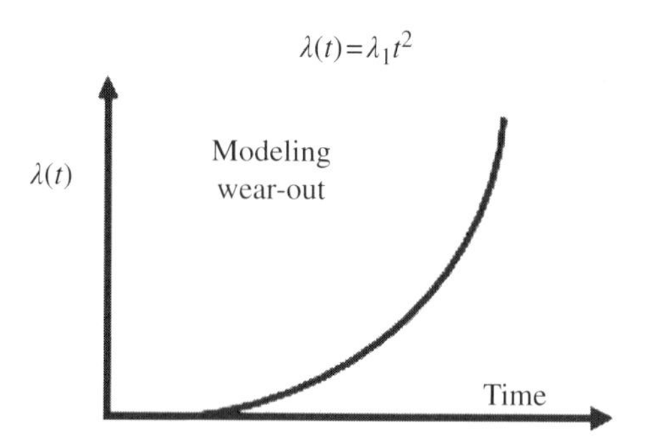

Figure A.2 Demonstrating the power law on the wear-out shape

A.4.2 *Weibull Distribution*

The Weibull distribution was named after Waloddi Weibull who first published it in 1951 [4]. As mentioned, reliability distributions model portions of the bathtub curve. Let's overview this a bit to simplify some of the mystery behind the Weibull function. If we were to try and say model the wear-out portion of the bathtub curve's shape in Figure A.1, what function would we select? We know it is non-linear so let's try a power law, for example $\lambda(t) = \lambda_1 t^2$ as shown in Figure A.2. This indeed gives a good shape.

We could therefore come up with a general model,

$$\lambda(t) = \lambda_1 t^Y.$$

This is essentially the AT&T (American Telephone and Telegraph) -equivalent Weibull model [5]. In that model, we note that if $Y = 0$ we are in the flat portion of the bathtub curve; if $Y < 0$, we would be in the infant mortality region; while if $Y > 1$ we are in the wear-out portion. The Weibull distribution is presented in a similar manner but, instead of Y, Weibull used essentially $Y = \beta - 1$ (see Equation (A.17)). The full two-parameter popular Weibull model is:

- Reliability function:

$$R(t) = \exp\left[-\left(\frac{t}{\alpha}\right)^{\beta}\right]. \tag{A.14}$$

- CDF (see Equation (A.2)):

$$F(t) = 1 - \exp\left[-\left(\frac{t}{\alpha}\right)^{\beta}\right] \quad \text{or} \quad \ln\left[\ln\frac{1}{1-F(t)}\right] = \beta\ln(t) - \beta\ln(\alpha). \tag{A.15}$$

- PDF (see Equation (A.3)):

$$f(t) = \frac{\beta}{\alpha^{\beta}} t^{\beta-1} \exp\left[-\left(\frac{t}{\alpha}\right)^{\beta}\right]. \tag{A.16}$$

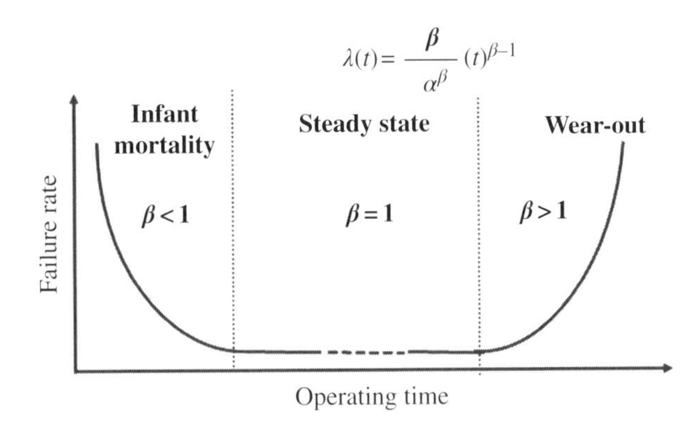

Figure A.3 Modeling the bathtub curve with the Weibull power law

- Failure rate (see Equation (A.4)):

$$\lambda(t) = \frac{\beta}{\alpha^{\beta}}(t)^{\beta-1}. \tag{A.17}$$

The Weibull model for the bathtub curve is shown in Figure A.3. We note that in Figure A.2, we shifted the wear-out axis to time zero. Alternatively, this can be accomplished by including a third time shift parameter. Such a three-parameter Weibull model is also used, but not detailed here. However it is essentially just a time shift. We note that any of the reliability distributions can have a time shift. For one reason or another, it is only customary to include it in the popular Weibull model.

The Weibull model is often considered the premier reliability model. This is because of the physical significance of β: the wear-out region is indicated by $\beta > 1$; the infant mortality region by $\beta < 1$; and steady state by $\beta = 1$, as illustrated in Figure A.3. Therefore reliability engineers tend to use it often; when life test data are fitted to the CDF, β is determined and the knowledge of where the data lie relative to the bathtub curve is very helpful in determining the product issues. The key is to remember that, although the Weibull distribution seems to have this advantage, the best distribution to use is the one that best fits the data.

Figure A.4 illustrate the behavior of the hazard rate for a number of different values of β, while Figures A.5 and A.6 present an idea of how the PDF and CDF functions behave for $\beta = 2$ (wear-out) and $\beta = 0.5$ (infant mortality) (see dfrsoft, module 2, row 115 [1]).

A.4.3 Normal (Gaussian) Distribution

The normal (or Gaussian) distribution is important. It has so many practical applications in statistics and is used in quality, reliability, economics, biology, physics, etc. The reader might recall that we discussed uses of the Gaussian distribution when we looked at white noise in Chapter 2, which is a Gaussian distribution. In terms of its use in reliability, it is primarily used in variable data. This differs from what we typically make use of when working in reliability with the Weibull or lognormal functions, which we primarily use to analyze catastrophic pass–fail data. One should note that the normal distribution function can be used to fit catastrophic data as well. However, it is atypical to find catastrophic data analyzed using

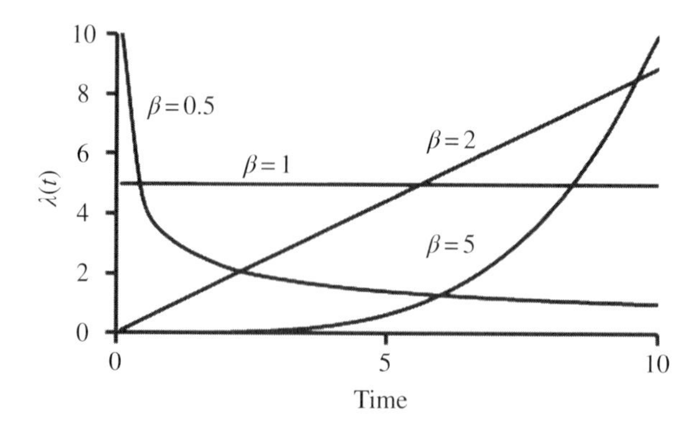

Figure A.4 Weibull hazard (failure) rate for different values of β [1]

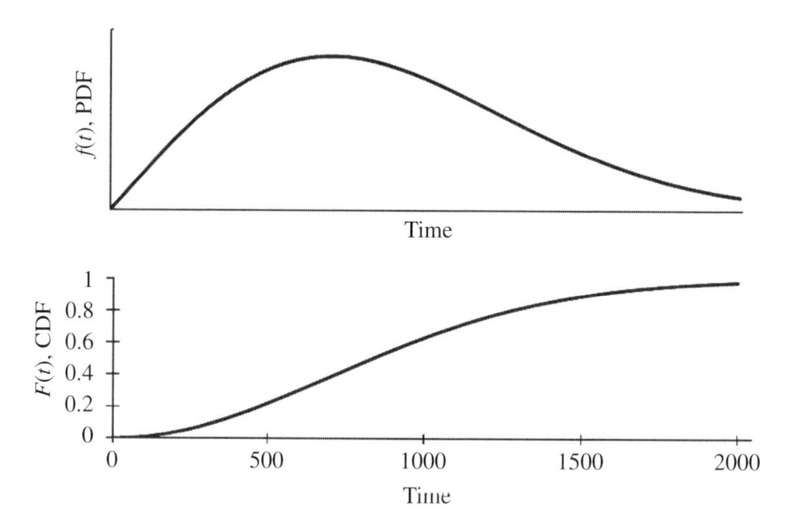

Figure A.5 Weibull shapes of PDF and CDF with $\beta = 2$ [1]

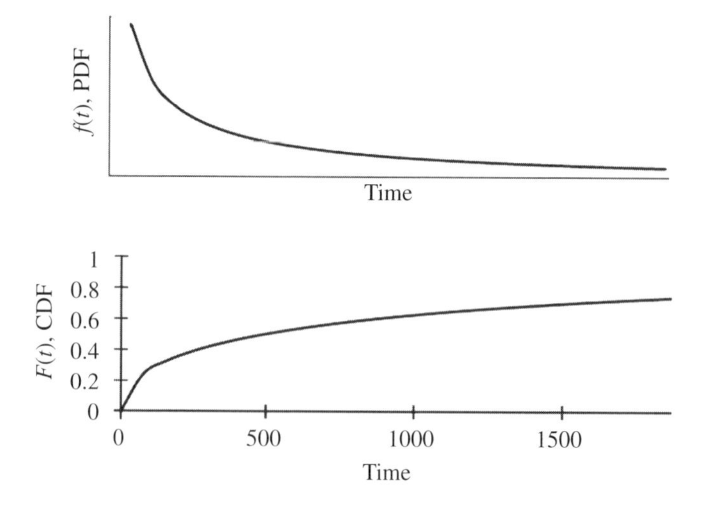

Figure A.6 Weibull shapes of PDF and CDF with $\beta = 0.5$ [1]

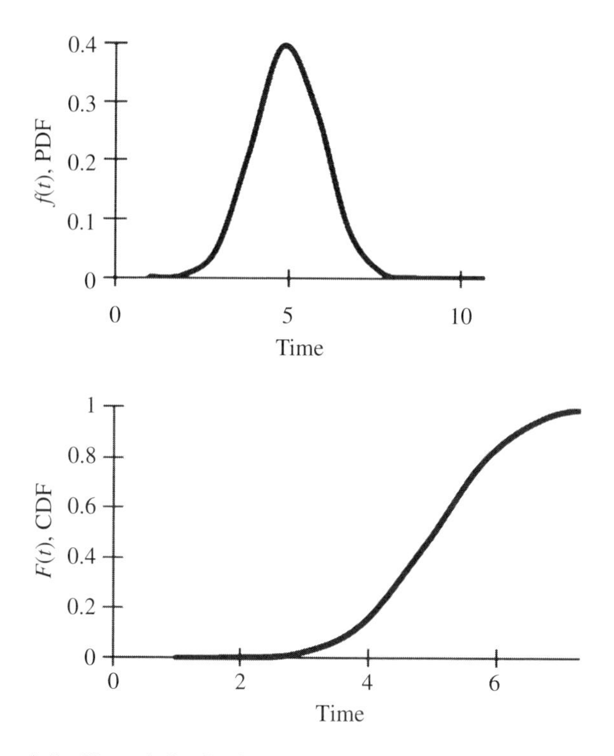

Figure A.7 Normal distribution shapes of PDF and CDF; $\mu = 5$, $\sigma = 1$ [1]

it. The popular PDF bell-shaped curve is probably the first thing one remembers when thinking of it, as shown in Figure A.7. The distribution is summarized below (this formulation will also help us to understand the lognormal distribution).

- PDF (see Equation (A.3)):

$$f(t) = \frac{1}{\sigma\sqrt{2\pi}} \exp\left[-\frac{1}{2}\left(\frac{x-\mu}{\sigma}\right)^2\right]. \tag{A.18}$$

- CDF approximation:

$$F(t) = \frac{1}{\sigma\sqrt{2\pi}} \int_{-\text{¥}}^{x} \exp\left[-\frac{1}{2}\left(\frac{x-\mu}{\sigma}\right)^2\right] = \frac{1}{2}\left\{1 + \operatorname{erf}\left[\left(\frac{x-\mu}{\sqrt{2\sigma}}\right)\right]\right\}. \tag{A.19}$$

- Population mean and variance:

$$\mu = \frac{\sum_{i=1}^{N} x_i}{N} \quad \text{and} \quad \sigma^2 = \frac{\sum_{i=1}^{N} (x_i - \mu)^2}{N}. \tag{A.20}$$

- Sample mean and variance:

$$\bar{x} = \frac{\sum_{i=1}^{n} x_i}{n} \quad \text{and} \quad s^2 = \frac{\sum_{i=1}^{n} (x_i - \bar{x})}{n-1}. \tag{A.21}$$

Figure A.7 illustrates the PDF and CDF functions for the normal distribution.

A.4.3.1 Example A.3: Power Amplifiers

Power amplifiers are distributed normally with a mean of 5 W and a variance of 1 W. What percent of the population is below 4 W?

The plot of the PDF and the CDF is given in Figure A.7 for these semiconductors. Note that those plots normally have the *X*-axis in time, but here we simply substitute watts. We can see from the CDF plot in Figure A.7 that the fraction below 4 W is about 15%. Using Equation A.19, the results is calculated as (see dfrsoft, module 5, row 23 [1]):

$$F(t) = \frac{1}{2}\left\{1 - \text{erf}\left[\left(\frac{4-5}{\sqrt{2}}\right)\right]\right\} = \frac{1}{2}[1 + \text{erf}(-0.707)] = 0.159$$

(note that $\text{erf}(-x) = -\text{erf}(x)$). If the specification limit for power amplifiers was 4 W or higher for shipments, we would lose about 15.9% of the population.

A.4.4 The Lognormal Reliability Function

The lognormal distribution is often used to model semiconductors. We noted in Chapter 9 that it has a lot of physical significance. That is, many parametric aging laws were described using the TAT models (see Chapters 6 and 7) that describe aging in log(time) dependence of key parameters. Creep was a good example. If we have a parametric threshold for failure, and the parameters are normally distributed and age in log(time) relative to the parametric threshold, then the failure rate is lognormal in time (this is detailed in Chapter 9). It is also reasonable to assume that catastrophic lognormal failure rates follow if microscopic aging is occurring in log(time) and the product spends most of its lifetime with this aging dependence. We also noted that many empirical aging power laws exist that are likely log(time), but can also be modeled as a power law rather than log(time) aging law (see Figure 6.3).

The lognormal distribution can be obtained similarly to the normal distribution functions by taking the logarithm of the time parameter. The results are as follows.

- PDF (see Equation (A.3)):

$$f(t) = \frac{1}{\sigma t\sqrt{2\pi}} \exp\left[-\frac{1}{2}\left(\frac{\ln t - \ln t_{50}}{\sigma}\right)^2\right]. \tag{A.22}$$

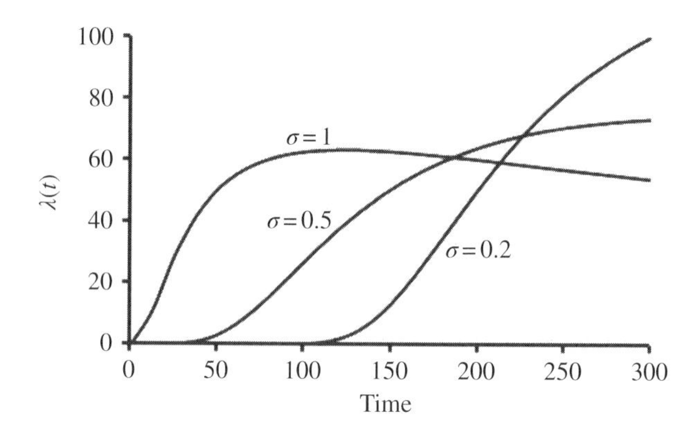

Figure A.8 Lognormal hazard (failure) rate for different σ values [1]

- CDF (see Equation (A.2)):

$$F(t) = \frac{1}{\sigma\sqrt{2\pi}}\int_0^t \frac{dx}{x}\exp\left[-\frac{1}{2}\left(\frac{\ln(x/x_{50})}{\sigma}\right)^2\right] = \frac{1}{2}\left[1+\operatorname{erf}\left(\frac{\ln(x/x_{50})}{\sqrt{2}\sigma}\right)\right]. \tag{A.23}$$

- PDF (see Equation (A.3)):

$$f(t) = \frac{1}{\sigma t\sqrt{2\pi}}\exp\left[-\frac{1}{2}\left(\frac{\ln t - \ln t_{50}}{\sigma}\right)^2\right]. \tag{A.24}$$

- Failure rate (see Equation (A.4)):

$$\lambda(t) = \frac{f(t)}{1-F(t)}. \tag{A.25}$$

- Median = $t_{50.}$ Shape parameter:

$$\sigma = \ln(t_{50}/t_{16}). \tag{A.26}$$

Figure A.8 illustrates the lognormal hazard rate for different shape parameters. We also illustrate this for the CDF and the PDF functions in Figure A.9 (see dfrsoft, module 2, row 155 [1]). In Section A.7 we provide an example comparing the Weibull to lognormal analysis on life test data.

A.5 Confidence Interval for Normal Parametric Analysis

The normal distribution is used a lot in quality and reliability analysis. Typical areas in industry include: Cpk analysis (design maturity testing); parametric analysis (process reliability testing); accelerated testing of parametric data; and parametric confidence.

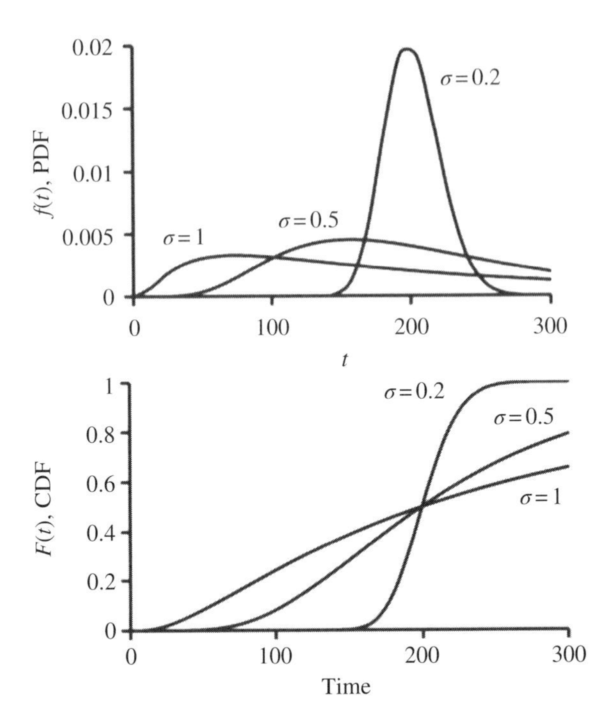

Figure A.9 Lognormal CDF and PDF for different σ values [1]

In industry, often we cannot determine the population mean and sigma as we are doing sampling. When sampling is involved we can estimate the population mean from the sample mean using the confidence interval

$$\bar{x} - z_{\alpha/2}\frac{\sigma}{\sqrt{n}} < \mu < \bar{x} + z_{\alpha/2}\frac{\sigma}{\sqrt{n}} \tag{A.27}$$

where $Z_{\sigma/2}$ is the Z value of a standard normal distribution leaving an area of $\alpha/2$ to the right of the Z value. The Z-value for any random variable x is $Z = (x-\mu)/\sigma$. For small samples when $n < 30$ normal population mean is approximated using the student t distribution, where μ is given by

$$\bar{x} - t_{\alpha/2}\frac{s}{\sqrt{n}} < \mu < \bar{x} + t_{\alpha/2}\frac{s}{\sqrt{n}} \tag{A.28}$$

where $t_{\alpha/2}$ is the t value with $v = n-1$ degrees of freedom, leaving an area of $\alpha/2$ to the right. The confidence equation is illustrated in the following example.

A.5.1 Example A.4: Power Amplifier Confidence Interval

In Section A.4.3.1, the power amplifiers were observed to have a mean of 5 W and sample sigma of 1 W. If this was determined for 40 amplifiers, what is the confidence around the population mean?

Analysis: We note that the sample size is larger than 30 units so that Equation (A.27) applies. For the 90% confidence interval $\alpha = 1 - 0.9 = 0.1$ and $\alpha/2 = 0.05$. For $Z_{0.05} = 1.64$, the confidence interval is (see dfrsoft, module 5, row 23 [1])

$$5-(1.64)\left(\frac{1}{\sqrt{40}}\right) < \mu < 5+(1.64)\left(\frac{1}{\sqrt{40}}\right)$$

which reduces to

$$4.73 < \mu < 5.27.$$

This is the 90% confidence interval about the mean. We are 90% confident that the mean falls within these limits. That is, if we took sample populations and measured means and sigma, 9 times out of 10 the results should fall within this range.

A.6 Central Limit Theorem and Cpk Analysis

The central limit theorem is important because it is a reason that many of our procedures such as SPC work well. It simply states that: the sampling distribution of the mean of any independent, random variable will be normal or nearly normal, if the sample size is large enough.

The next question is: how large is large enough? Often people default to a size $n \geq 30$ as this size is considered to have reduced fluctuations in the value of the variance. If the population is reasonably well behaved without being too skewed, this is a good rule of thumb. Otherwise, some statisticians like to see a minimum of 40 devices.

A.6.1 Cpk Analysis

Industry uses Cpk as a major metric for quality control. It is a value that can also be related to yield (see Table A.2). It is also used in reliability qualification testing. In that application, samples are selected and key parameters are assessed using the Cpk index as a measure before and after qualification; the Cpk values are then compared. In order to use the Cpk method, the parameter being measured should be normally distributed. Figure A.10 illustrates the use of the Cpk index with key values.

The Cpk upper and lower values are given by:

$$\text{Cpk}_\text{L} = \frac{(\bar{X} - \text{LSL})}{3s} \quad \text{and} \quad \text{Cpk}_\text{U} = \frac{(\text{USL} - \bar{X})}{3s} \tag{A.29}$$

where LSL and USL stand for lower and upper specification limit, respectively.

Here we select the worst-case minimum Cpk value to characterize the sample.

A.6.2 Example A.5: Cpk and Yield for the Power Amplifiers

We can find the Cpk for the power amplifiers and the corresponding yield to illustrate how Cpk works (see dfrsoft, module 21, row 10 [1]). Consider the power amplifiers example where the mean was 5 W; we can use 4 W as the LSL and 7 W as the USL, so that

Table A.2 Relationship between Cpk index and yield [1]

Cpk value	Two-sided (PPM)	Two-sided normal percent	One-sided (PPM)	One-sided normal percent
2.00	0.002	99.9999998	0.001	99.99999990
1.667	0.6	99.99994	0.3	99.99994
1.50	6.8	99.99932	3.4	99.99966
1.333	63	99.994	32	99.997
1.166	465	99.95	233	99.98
1.00	2700	99.73	1350	99.87
0.833	12 419	98.76	6210	98.76
0.667	45 500	95.45	22 750	97.73
0.500	133 615	86.64	66 807	93.32
0.333	317 311	68.27	158 655	84.13

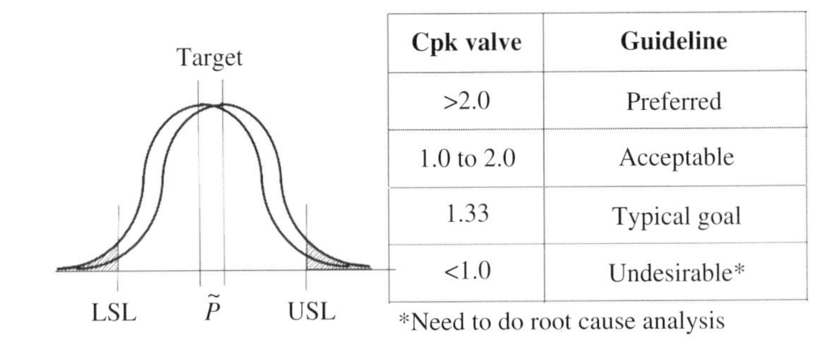

Figure A.10 Cpk analysis

$$\text{USL} - \text{Mean} = 2\,\text{W},\ \text{Mean} - \text{LSL} = 1\,\text{W}.$$

Therefore, the Mean – LSL is the minimum value. Since $\sigma = 1$, the sample Cpk index is given by Cpk_{LSL}, defined

$$\text{Cpk}_{\text{LSL}} = 1/(3 \times 1) = 0.33.$$

From Figure A.10, this is an undesirable Cpk value and likely requires some design improvements. We note from Section A.4.3.1 for this USL that about 16% of the product is out of specification (also this is shown in Table A.2).

It is instructive to look at the Cpk for the USL as well, this is

$$\text{Cpk}_{\text{USL}} = 2/(3 \times 1) = 0.66.$$

From Table A.2 we see that about 2.3% is out of specification on the high end. The total product out of specification is 2.3% + 16% = 18.2%. This illustrates the concept of Cpk and how it relates to yield.

A.7 Catastrophic Analysis

We have looked at a number of uses in the area of parametric analysis such as Cpk and yield. Another excellent area that is a bit outside the scope of this chapter is parametric reliability analysis. The reader is referred to dfrsoft, module 16 [1] for more information in this area.

In this section we will detail catastrophic failure analysis specifically for life testing when there are say greater than 5% failures observed. When we have few failures, the underlying distribution is likely a constant failure rate and we are testing in the steady-state portion of the bathtub curve. In this case, we will use the methods in Section A.8 (also see Special Topics Sections B.1.3 and B.6). When we have a large number of failures on life test, the failure population is likely in the infant mortality or wear-out region of the bathtub curve. In order to analyze the data, we must look specifically at a particular failure mechanism. That is, when doing life test analysis using Weibull or lognormal distributions, each failure mechanism has its own characteristic life and work path to failure affecting the statistics. We will address mixed modes in Section A.7.3 as well. The process to analyze life test data is as follows.

Arrange the failures in increasing time of occurrence. Since we will be using the CDF to fit the data, the most common unbiased plotting method for cumulative i failures out of a total of n devices is the median ranked plotting position for the CDF:

$$F_i = (i - 0.3)/(n + 0.4).$$

The reader may wonder why we do not use a simple plotting position such as i/n. The main reason is that it is biased in that we never plot the zero point but would be able using this method to plot the 100 percentile point. This is then considered biased, while a plotting position such as $(i - 0.3)/(n + 0.4)$ is considered unbiased in this regard.

A.7.1 Censored Data

Typically in reliability testing we may end the test before all the failures have occurred and/or we may not know the exact failure time, and so forth. We therefore have what is called censored data. Such data are said to be *censored on the right* (failure time $> t_0$). A failure time known only to be before a certain time is said to be *censored on the left* (failure time $< t_0$). However, if a failure time is known to within an interval, when it is not continuously monitored, it is said to be *interval censored* ($t_0 <$ failure time $< t_1$). If all units are started together on test and the data are analyzed before all units fail, the data are singly censored. Data are *multiply censored* if units have different running times, intermixed with the failure times. Time-censored data are also called Type 1 censored; this is the most common type of censored data. There are a number of methods used in data analysis for singly and multiply censored data [2, 3].

A.7.2 Example A.6: Weibull and Lognormal Analysis of Semiconductors

In Table A.3 we list life test data for an accelerated semiconductor test. Devices were put on test at 225°C under bias condition with a junction rise of 10°C. Find the estimated MTBF at 80°C use conditions. Assume an Arrhenius acceleration model with 0.7 eV. What is the estimated percent of the population that will fail in 15 years?

Table A.3 Life test data arranged for plotting

Cumulative failures, i	No. failures	Cumulative failures, i	Plotting position $F_s = (i - 0.3)/(n + 0.4)$ $N = 15$	Failure times at 200°C (h)
1	1	1	4.55	500
2	1	2	11.04	650
3	1	3	17.53	725
4	1	4	24.03	900
5	2	6	37.01	1000
6	3	9	56.49	1200
7	2	11	69.48	1500

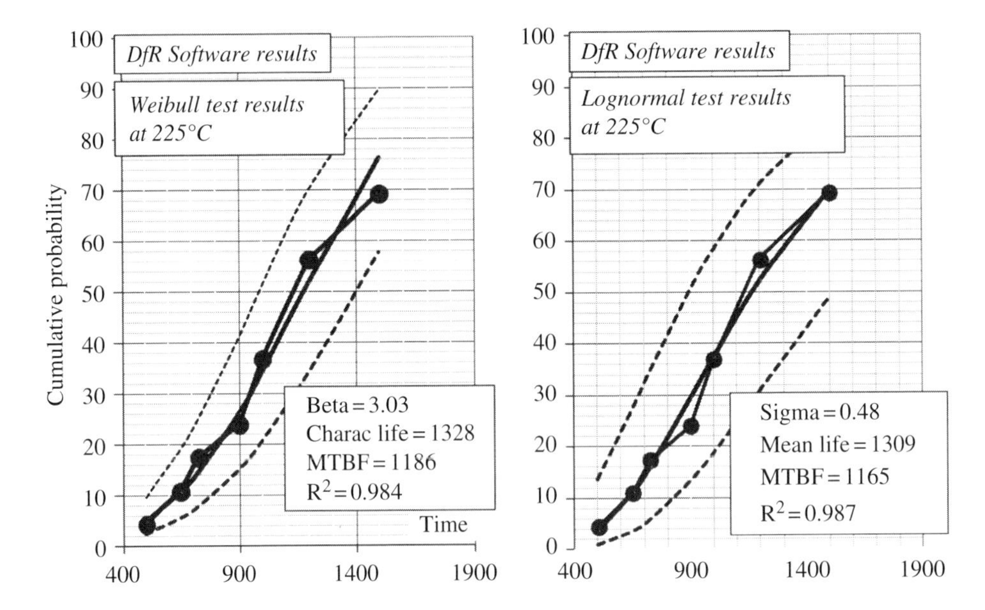

Figure A.11 Life test: (a) Weibull analysis compared to (b) lognormal analysis test at 200°C [1]

Analysis: (See dfrsoft, module 2, row 13 [1].) We have arranged the data in increasing failure times and plot the median ranked position. We are now able to use a linearized form of the Weibull Equation (A.15) to fit the data and find the Weibull parameters. Similarly, we do this for the lognormal distribution. Using the linearized version of these CDFs, the multiple regression analysis best fit is obtained.

The results are displayed in Figure A.11 for both the Weibull and lognormal (see dfrsoft, module 2, row 13, Col T and AA [1]). First note that $\beta > 1$ for the Weibull, which indicates that we are in the wear-out portion of the bathtub curve. This tells us that we do not expect poor reliability performance. We note the regression coefficients are 0.984 for the Weibull and 0.987 for lognormal, that is, a slightly better fit for the lognormal analysis.

Next we would like to project this result to use condition. The junction test temperature is 225°C + 10°C = 235°C, and the actual junction use condition is 80°C + 10°C = 90°C. Using the Arrhenius acceleration factor (see Equation (5.20); Section B.1.2.1) with 0.7 eV we obtain an

Table A.4 Field data and the renormalized groups

Failure time (months)	Number of units that fail at the time	Main population $(i-0.3)/(n+0.4)$	Lower distribution	Upper distribution
			Total units = $17(i-0.3)/(n+0.4)$	Fail = 5, Susp = $18(i-0.3)/(n+0.4)$
1	3	6.7	15.9	
2	2	11.6	27.7	
3	2	16.6	39.5	
4	1	19.1	45.4	
5	1	21.5	51.3	
6	1	24.0	57.2	
7	1	26.5	63.1	
8	1	29.0	69.0	
10	1	31.4	74.8	
13	1	33.9	80.7	
17	1	36.4	86.6	
21	1	38.9	92.5	
24	1	41.3	98.4	
27	1	43.8		3.1
42	1	46.3		7.4
51	1	48.8		11.7
60	1	51.2		15.9
69	1	53.7		20.2

estimated acceleration factor of 591. Then the Weibull MTBF = 591 × 1186 = 700 926 h and the characteristic life is 1328 × 591 = 784 848 h. The fraction that will fail in 15 years (= 131 400 h) is, according to Equation (A.15):

$$F(15\,\text{years}) = 1 - \exp\left[-\left(\frac{131\,400}{784\,848}\right)^{3.03}\right] = 0.0044 = 0.44\%.$$

A.7.3 Example A.7: Mixed Modal Analysis Inflection Point Method

If a distribution is made up of two failure mechanisms, it is difficult to see the two modes if the data had been plotted without separating out failure mechanisms first. This can occur if both modes fail in the same time frame throughout the test. Thus, the only way to analyze the data is by doing failure analysis on the product and determining the actual failure mechanism for each failure observed.

However, in life testing a subpopulation and main population are often observed, occurring at two distinct time frames. The subpopulation (sometimes called "freaks" or infant mortality failures) show up at early test times compared to the main failure population.

This also commonly occurs in field data. Field data are presented in Table A.4 (see dfrsoft, module 2, row 13, Col. B and K [1]). This characteristic for the data in Table A.4 can be observed by an inflection point in the life test data as shown in Figure A.12 at the 42% point.

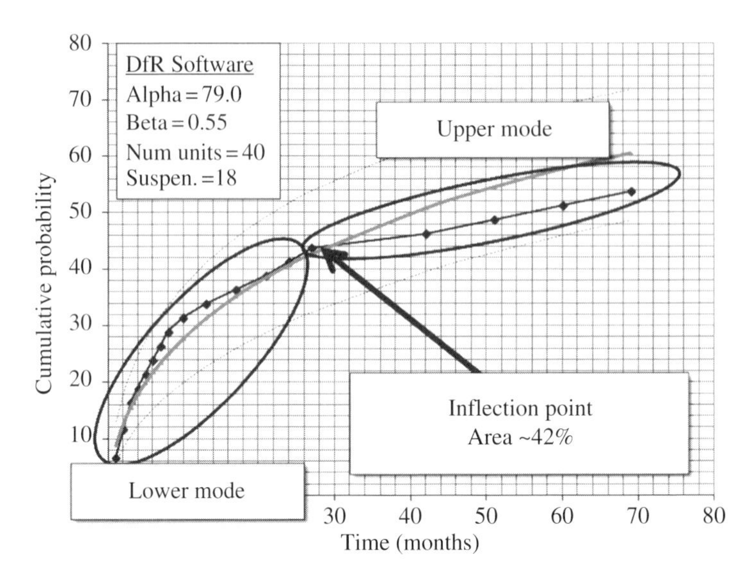

Figure A.12 Field data (Table A.4) displaying inflection point as sub and main populations [1]

This behavior can occur from the same failure mode with two failure mechanisms; early failure types could be mechanical failures while later failures could be, say, electrical. Recognizing this, one sees a distinct separation at an inflection point (sometimes looking like S-shaped data, other times just a simple inflection) in the cumulative probability plot as displayed in Figure A.12, with inflection point at 42%. This divides the total population into lower (early time) 42% subpopulation and main or upper (later time) population (58%) groups. The field data for Figure A.12 in Table A.4 (see dfrsoft, module 2, row 13, Col. T [1]) are obtained from 40 units with 22 failures observed and 18 suspensions (units that have not failed at the last observation time). The observed failure times are listed in the table. Here we assume that all the suspensions are part of the upper population.

The inflection point analysis method is fairly straightforward. The lower and upper populations are simply renormalized by treating each separately above and below the inflection point. This is illustrated in Table A.4 using a Weibull analysis. First we have provided in columns 1 and 2 the failure times and the corresponding number of units failing at that time (interval). In column 3, we list the cumulative percent failure plotting position prior to separating out the population which is plotted and fitted to a Weibull distribution in Figure A.12. Then this graph provides us with the observed inflection point. If this point is not visible, it is likely that there are not two distinct populations. Often one is tempted to do multimodal analysis even when the inflection point is not there. Without some knowledge of an early failure population either statistically (by seeing the inflection point) or from field data analysis of the failure mechanisms on failed units, it is likely that such an analysis will not be justified. However, as in this case when there is an inflection point, then one renormalizes the lower population as in column 4 and similarly for the upper population in column 5. One can then proceed and do a Weibull analysis or, if need be, a lognormal analysis. The results of the main, upper, and lower population are displayed in Figure A.13 (see dfrsoft, module 2, row 55, Col. AU or use the hyperlink [1]).

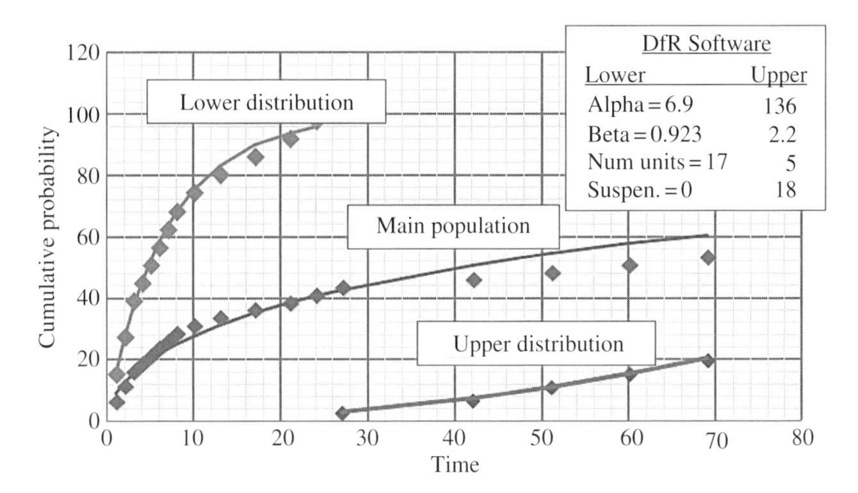

Figure A.13 Separating out the lower and upper distributions by the inflection point method [1]

Note that in Figure A.13 we see that the Weibull results indicate that the lower population indeed has a $\beta < 1$ (0.92), indicating that it is an infant mortality problem; the upper population has a $\beta > 1$ (2.2), indicating a wear-out failure mode.

A.8 Reliability Objectives and Confidence Testing

We described catastrophic analysis in Section A.7, where we observed a reasonable percent of the sample population on test failing. We are now interested in demonstration testing where we have few failures if any (less than a few percent of the population failing). To this end, we need to do what called statistical confidence demonstration testing.

There are two kinds of confidence: engineering confidence; and statistical confidence.

In engineering confidence we use judgment. For example, based on your experience you are confident that the experiment shows you have a reliable product. Often this may be a valid view. We use engineering judgment all the time in design and in testing as well as in manufacturing products. Without engineering confidence, technology might still be in the dark ages.

However, we often need to go a step further and require more than engineering confidence; we would therefore like to have a certain measure of statistical confidence. Statistical confidence plays a big role not only in product reliability but also in areas such as the medical industry. Statistical confidence in this section implies that we will select a statistically significant sample size, and test the device for a statistically significant period of time to assess the reliability of the product under appropriate stresses. This is in fact the way qualification testing is done. Customers often want to see the data. They want to buy a product that has been statistically verified to be reliable. For example, depending on the complexity of the item being qualified, the test objective might be

- for small parts, ICs, $\bar{\lambda} = 0.4$FMH (0.34%/year);
- for complex hybrids, $\bar{\lambda} = 4$FMH (4.33%/year); or
- for complex assemblies, $\bar{\lambda} = 10$FMH (8.4%/year).

A.8.1 Chi-Squared Confidence Test Planning for Few Failures: The Exponential Case

When we have few failures in an accelerated life test, we essentially must be in that portion of the bathtub curve called steady state. As mentioned in Section A.4.1, the steady-state portion of the bathtub curve has a constant failure rate is modeled by the exponential distribution.

The problem is how best to design a test for few failures. Reliability engineers often use what is called the chi-squared distribution in test design for this situation. It is shown in Section A.8.1.1 that when the failure rates are exponentially distributed the chi-squared distribution can be used in design of reliability tests. This is the option that we will describe here. The chi-squared equation for testing [2, 3] is

$$\bar{\lambda} = \frac{\chi^2\{\gamma; 2y+2\}}{2N(\text{AF}t)} \tag{A.30}$$

where $\bar{\lambda}$ = the upper bound on the failure rate; N is test sample size; AF is the acceleration factor; t is the test time; χ^2 is the chi-squared statistic (found in tables); γ is the confidence level (typically 60, 80, or 90%); and y is the planned or observed number of test failures.

A.8.1.1 Chi-Squared Validity

Here we provide a bit more information on the validity of the chi-squared method. In statistics, the chi-squared equation is also written for simple statistics when we wish to know the upper bound fractional failure portion P as

$$\bar{P} = \frac{\chi^2\{\gamma; 2y+2\}}{2N}. \tag{A.31}$$

This is presented in statistics as valid for $y/N < 10\%$. That is, we now know the upper bound fractional failure. Now we recall in Section A.4.1.1 that $F(t)$ for the exponential distribution can be approximated by $F(t) \approx \lambda t$ for $\lambda t << 1$. This is only valid for the exponential distribution with its constant failure rate. Then:

$$\bar{\lambda} = \bar{P}/t = \frac{\chi^2\{\gamma; 2y+2\}}{2Nt} \quad \text{where } t \rightarrow \text{AF} \times t. \tag{A.32}$$

In an accelerated test we simply replace t by AF × t, yielding Equation (A.30).

We now have roughly two restrictions for the accuracy of Equation (A.30): $y/N < 10\%$ and $\lambda t << 1$. Note that y/N is not much of a restriction for accelerated testing as N has essentially been replaced by $y/(N\ \text{AF}\ t) < 10\%$. A simple way to think of the restriction $\lambda t << 1$ is to rewrite it as $t <<$ MTBF.

An alternate method for exact accelerated test planning is called the binomial exact sampling method (see dfrsoft, module 8, row 145 [1]). This is beyond the scope of this chapter.

A.8.2 Example A.8: Chi-Squared Accelerated Test Plan

As an example, consider an accelerated test to be planned for a complex hybrid with the objective for hybrids given in Section A.8 of 4 FMH. We would like to estimate the sample size for the following test: $\bar{\lambda} = 4\,\text{FMH}$; test time = 1000 h; AF = 85. Allow for 1 failure to occur. $\gamma = 90\%$. What is N?

In this example, we plan for 1 failure to perhaps occur, the confidence is 90%, and we wish to find the sample size N. This is basically a plug-in problem. The χ^2 value may be found in statistical tables where:

$$\chi^2\{\gamma; 2y+2\} = \chi^2\{90\%; 2\times 1+2\} = \chi^2\{90\%; 4\} = 7.78.$$

Then the sample size is (see dfrsoft, module 8, row 15, col. E [1])

$$N = \frac{\chi^2\{\gamma; 2y+2\}}{2\bar{\lambda}(\text{AF}t)} = \frac{7.78}{2\times 4\,\text{FHM}\times(85\times 1000)} = 11.$$

Note that the point estimate for the failure rate is $\lambda = 1.07\,\text{FMH}$.

We are 90% confident that if we perform this test with no more than 1 failure occurring, that the failure rate will be no higher than 4 FMH. That is, we will observe no higher than this failure rate 90 times out of 100.

A.9 Comprehensive Accelerated Test Planning

We see how important accelerated testing is for being able to demonstrate a failure rate objective in a reasonable test period. In Special Topics B we describe numerous accelerated test examples based on the physics of this book. Such models are necessary to estimate the effects of raising the level of the appropriate stress to accelerate a potential device failure mode and effectively compress time. Thus, estimating time compression strongly influences test planning. Once the overall acceleration factor is estimated, tests can be properly planned.

Multiple accelerated testing is also done for sophisticated planning. Table A.5 illustrates an optimized test plan. We will detail the results in Special Topics Section B.6.

Table A.5 Multiple stress accelerated test to demonstrate 1 FMH [1]

Stress	Example number	Acceleration factor	Test time	FMH	Minimum sample size
High-temperature operating life (HTOL)	B.1	114	1000 h	0.233	87
Temperature–humidity bias (THB)	B.5	295	1000 h	0.144	54
Temperature cycle (TC)	B.6	25	200 cycles	0.226	85
Total				0.6	226

References

[1] Author's software, download at http://www.dfrsoft.com (see Section A.1.1 for details) to follow along with examples.

[2] O'Connor, P. and Kleyner, A. (2012) *Practical Reliability Engineering*, 5th edn, John Wiley & Sons, Ltd, Chichester.

[3] Feinberg, A., Crow, D. (eds) *Design for Reliability*, M/A-COM 2000, CRC Press, Boca Raton, 2001.

[4] Weibull, W. (1951) A statistical distribution function of wide applicability. *Journal of Applied Mechanics*, **18**, 293–297.

[5] Klinger, D.J., Nakada, Y. and Menendez, M.A. (1990) *AT&T Reliability Manual*. Van Nostrand Reinhold, New York.

Special Topics B

Applications to Accelerated Testing

B.1 Introduction

In this section we provide some examples of how to use the acceleration models detailed in Chapter 4 and elsewhere in this book.

Accelerated testing is the cornerstone of reliability testing by companies all over the world. There are a number of reliability conferences where the majority of the papers pertain to the subject of accelerated testing done on products in industry. The reader is referred to two of the most popular conferences called *Accelerated Stress Testing and Reliability* (ASTR) conference and *Reliability and Maintainability Symposium* (RAMS). In many industries, "Failure is not an Option"; this is in fact a slogan often coined by NASA. It fact NASA engineers make up the largest group of attendees at these conference every year. In terms of thermodynamic reliability, entropy damage in systems is a big concern and what better way to look for potential entropy problems than by doing accelerated testing. Tests are typically divided up into five main categories related to the popular stresses used in industry: high temperature; thermal cycle; temperature–humidity; salt fog; and shock and vibrations corrosion testing. These tests are exemplified in Figure B.1.

We note that high-temperature accelerated testing is often used to look at issues of intermetallics and diffusion as well as thermomechanical failure mechanisms (that can include wire bond integrity, Kirkendall voiding, ohmic contact issues, electromigration) and semiconductor junction issues such as metal oxide semiconductor (MOS) gate wear-out and so forth. Heat is often an enemy as it increases the devices entropy. As in semiconductor testing, devices are often biased to include the high-temperature junction rise.

Thermodynamic Degradation Science: Physics of Failure, Accelerated Testing, Fatigue, and Reliability Applications, First Edition. Alec Feinberg.

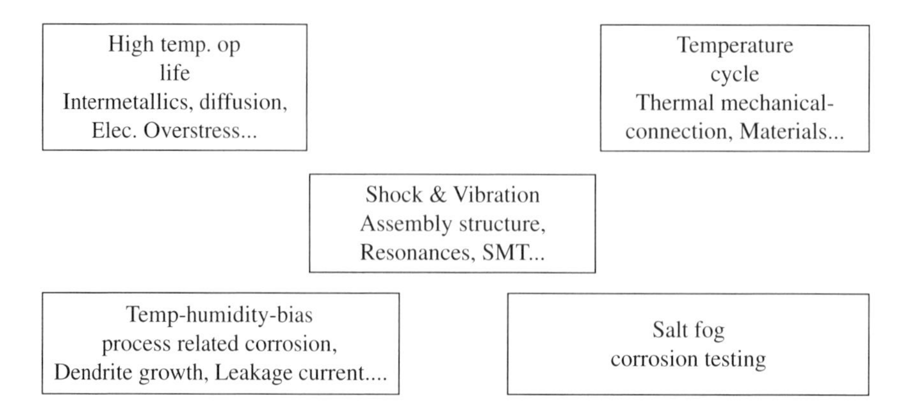

Figure B.1 Main accelerated stresses and associated common failure issues

Temperature cycle accelerated testing is often used to look at expansion–contraction temperature coefficient of thermal expansion (TCE) mismatch issues. For example, one of the most common issues in electronic circuit boards is the TCE mismatch between large solder components and the printed circuit board. Ceramic ball grid arrays (BGAs) for example in contact to a common FR-4 board can have a TCE mismatch of 11.5 ppm/°C. Such mismatches cause stress which, in combination with a large-size BGA, can cause early life failure.

Temperature–humidity–bias testing is also a very common accelerated test. Harsh tropic environments with high humidity present challenges for electronic components in the field. This test looks for moisture-related failure mechanisms such as silver migration from silver-filled epoxies, electrical breakdown, plastic IC moisture ingress leading to complex solder reflow expansion package cracking issues, and so forth.

Shock and/or vibration are important tests as products in the field experience minimally transportation vibration in shipping. As well some products are in harsh shock and vibration environments such as automotives and aircraft. Typically, shock testing is not accelerated as it can often be done on a repetitive electrodynamic shaker where numerous shocks can be applied to equate to the actual product life. However, vibration testing needs to be accelerated.

Finally, corrosion testing can be important. It is estimated that corrosion problems have annually cost up to $350 billion in the US alone, of which 40% could likely be prevented. Accelerated testing (such as salt fog or humidity testing) can play a big part in understanding potential product corrosion issues.

B.1.1 Reliability and Accelerated Testing Software to Aid the Reader

It is helpful to use software in understanding the reliability and accelerated testing analysis described here and in Special Topics A. Free reliability statistics and accelerated testing software is available to aid the reader at the author's website, www.dfrsoft.com, to try for all the examples in Special Topics A and B [1]. This chapter provides reference to the module and row number that the reader can go to in the software to follow with the examples, for example (dfrsoft, module 2, row 13). The software is in friendly Excel format with reference modules (see menu) that includes help in the following areas: reliability conversion; reliability plotting

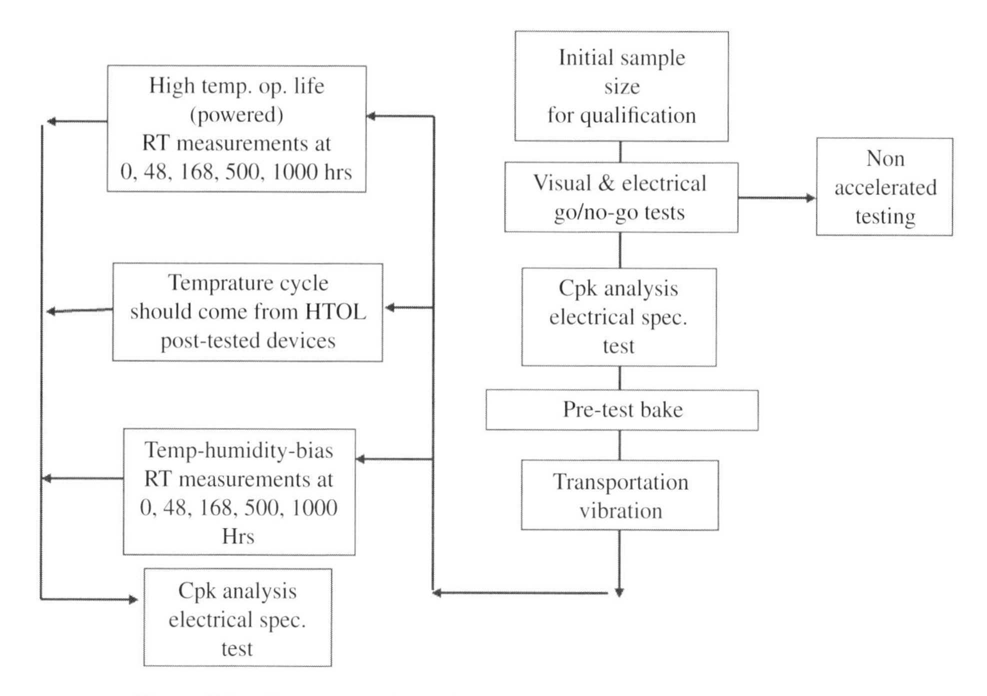

Figure B.2 Common accelerated qualification test plan used in industry [1]

(Weibull, lognormal, …); system reliability; distributions; field returns; acceleration factors; test plans; environmental profiling; reliability growth; reliability predictions; parametric reliability; availability and sparing; derating; physics of failure; Cpk analysis; lot sampling; SPC; thermal analysis; shock and vibration; corrosion; and design of experiment.

Testing for a product can often include multiple combinations of these tests [2, 3]. Figure B.2 illustrates a common test flow that one might use for multiple testing on a product (dfrsoft, module 15, row 207 [1]). We see that testing exposes the product flow to all stresses. Depending on the target failure modes if known, one can focus more on one stress than the others. Typically, the key stress may not be known. Most of the time, qualification testing is done using accelerated testing at a certain level of confidence. This allows a target reliability goal to be proven for the qualification test to within a certain upper confidence level of the failure rate. This will be discussed in the next sections (also see Section B.6.1).

B.1.2 Using the Arrhenius Acceleration Model for Temperature

Often, it is assumed that the dominant thermally accelerated failure mechanisms will follow the classical Arrhenius relationship described by Equation (5.20). The basic acceleration factor model is repeated here for convenience:

$$\mathrm{AF} = \frac{\tau_2}{\tau_1} = \frac{\mathrm{Rate}(T_1)}{\mathrm{Rate}(T_2)} = \exp\left\{ -\frac{E_a}{k_{\mathrm{B}}}\left(\frac{1}{T_1} - \frac{1}{T_2}\right)\right\} \tag{B.1}$$

where $k_B = 8.6173 \times 10^{-5}\,\text{eV/K}$ is Boltzmann's constant. Boltzmann originally proposed a similar relation as a probability distribution (see Section 6.3). There are a number of ways this model is used. In degradation, we can think of it in terms of defect activation. In this view, there is a certain probability of a defect surmounting the free energy barrier (see Figure 6.1). When enough defects are thermally activated over time, catastrophic failure results, the system's entropy is maximum (maximum disorder) and aging stops as the system is in its lowest free energy state, a state of final equilibrium. Note that since k_B has units of degrees Kelvin, so must the temperatures T_1 and T_2. Although we found this for corrosion, this relationship is relevant to many thermally activated processes. The rate can be written as

$$\text{Rate}(T) = \nu \exp\left\{-\frac{E_a}{k_B T}\right\} \tag{B.2}$$

where ν is a rate constant. We note that E_a can be thought of as the free energy barrier height. The barrier height is a property of the system's free energy related to the type of material (see Chapter 6 for more details). The expression is often written in linear $Y = ax + b$ form so that the time to failure can be plotted for regression analysis as:

$$\ln(\text{MTTF}) = \text{const} + \frac{E_a}{k_B}\left(\frac{1}{T}\right) \tag{B.3}$$

where the constant is simply $\ln(1/\nu)$.

Each failure mechanism that is thermally activated has associated with it an activation energy E_a. A number of different failure mechanisms and values for E_a can be found in Table 6.1, with general ranges often cited in the literature. We note that the higher the temperature, the more likely is the probability of a defect being excited and jumping over the free energy barrier. In a sense the barrier is a relative minimum of the free energy (see Figure 6.1), not the true minimum. However, when enough defects surmount the barrier and catastrophic failure occurs, the free energy is then at its true minimum.

In practice, when trying to estimate acceleration factors without knowing the activation energy value for each potential failure mechanism, a conservative value of E_a is used. For example, 0.7 eV is typical for IC failure mechanisms and appears to be somewhat of an industry standard when conservatively estimating test times (see Table 6.1).

B.1.2.1 Example B.1: Using the Arrhenius Acceleration Model

Consider a semiconductor device with a thermal resistance of 20°C/W (junction to ambient). The device is operated at 0.5 W. It is typically operated at 30°C and is tested at 100°C. What is the required test time to simulate 10 years of life in reliability testing? Assume an activation energy of 0.7 eV for the failure mechanism.

Analysis: First we find the power dissipated in the device at 0.5 W × 20°C/W. The junction rise is 10°C over ambient.

Then the field use condition is $T_{use} = 10°\text{C} + 30°\text{C} = 40°\text{C} = 273.15 + 40 = 313.15\,\text{K}$.

The stress temperature condition is $T_{stress} = 10°\text{C} + 100°\text{C} = +110°\text{C} = 383.15\,\text{K}$.

From Equation (B.1), the acceleration factor is (dfrsoft, module 7, row 13 [1])

$$AF_T = \exp\{(0.7\,\text{eV}/8.6173\times10^{-5}\,\text{eV/K})\times[1/(313.15)-1/(383.15)\,\text{K}]\} = 114.$$

The test time to simulate 10 years of life (87 600 h) is

$$\text{Test time} = \text{Lifetime}/AF_T = 87\,600/114 = 768\,\text{h}.$$

In the above we assumed an activation energy of 0.7 eV. One might ask: what if we wish to find this value? In process reliability for example, one can test multiple devices in groups at high temperature. Then the mean time to failure (MTTF) for each temperature group is assessed and the result can be plotted according to Equation (B.3) to find the activation energy for the failure mechanism associated with the test item. If for example only two temperatures are used, then we can solve Equation (B.1) directly for the activation energy which yields:

$$E_a = k_B \frac{\ln(\text{MTTF}_2/\text{MTTF}_1)}{(1/T_2 - 1/T_1)} \tag{B.4}$$

and we substitute the numbers. A more instructive problem is given below.

B.1.3 Example B.2: Estimating the Activation Energy

A process reliability study was performed on semiconductors at six different temperatures. The results yielded the following MTTFs.

Find the activation energy for the process and estimate the MTTF at a field use condition of 50°C.

Using (dfrsoft, module 7, row 99 [1]) Equation (B.3) we can plot ln(MTTF) against $1/T$. Then using a linear regression analysis, we can find the slope of the line. The slope should yield a value of E_a/k_B. The graphical result is presented in Figure B.3 (see also Table B.1; dfrsoft, module 7, row 139 [1]).

The slope yields an activation energy of 0.692 eV. Using this we can project the MTTF at 50°C, which will yield an MTTF of 464 909 h.

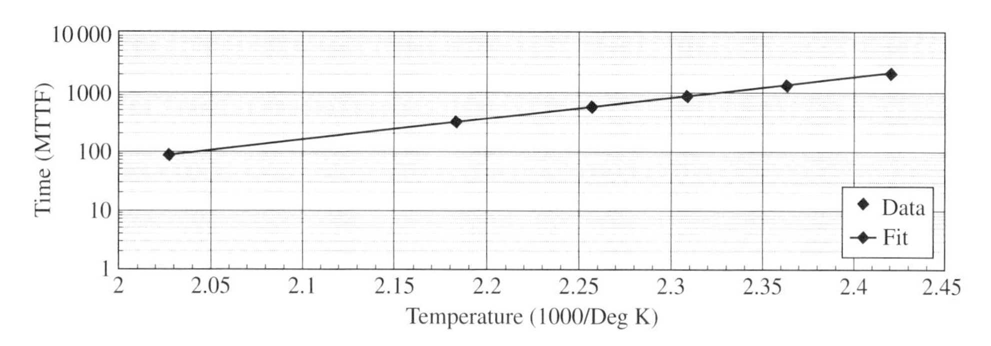

Figure B.3 Arrhenius plot of data given in Table B.1 Courtesy: DfR Software company

Table B.1 MTTF observed

Temperature (°C)	MTTF
140	2000
150	1400
160	850
170	550
185	300
220	90

B.1.4 Example B.3: Estimating Mean Time to Failure from Life Test

In Section B.1.2.1 what is the upper 90% confidence bound on the failure rate if 25 devices are tested and only 1 failure is observed? We use the method adopted in Section A.8. The chi-squared value is:

$$\chi^2\{\gamma;2y+2\}=\chi^2\{90\%;4\}=7.78.$$

The value $\chi^2(0.9,4)=7.78$ may be found in statistical tables. The device hours is $25\times 768\,\text{h}\times 114=2\,188\,800$ device hours. Assuming a constant failure rate, the 90% upper bound on the failure rate is (dfrsoft, module 8, row 11, col. m [1]):

$$\bar{\lambda}=\frac{7.78}{2\times 2\ 188\ 800}=0.00000177\ \text{failures/h}=1.77\ \text{FMH}.$$

The MTTF is $1/\lambda=218\ 800$ h, which is a lower bound 90% confidence limit on the MTTF. Note the point estimate is 0.46 FMH = 1 failure/2 188 800 device hours.

B.2 Power Law Acceleration Factors

Chapter 4 presented numerous acceleration factors; we can quickly summarize a few here for convenience. The acceleration factor found for creep in Equation (4.16) was

$$\text{AF}_{\text{creep}}=\left(\frac{\tau_2}{\tau_1}\right)=e^{-E_a/k_B\left(\frac{1}{T_1}-\frac{1}{T_2}\right)}\left(\frac{\sigma_1}{\sigma_2}\right)^K \tag{B.5}$$

and its linear relationship for the time to failure was noted in Equation (4.18) as

$$\ln(t_{\text{failure}})=C+\frac{E_a}{k_B T}-K\ln(\sigma). \tag{B.6}$$

We noted that the acceleration factor for wear, while not a power function, can be thought of as a power law with a power factor of 1. This was given in Equation (4.23) as

$$\mathrm{AF_{wear}}(2,1) = \frac{\tau_1}{\tau_2} = \left[\frac{\left(P_{W2} \dfrac{\nu_2}{A_2} \right)}{\left(P_{W1} \dfrac{\nu_1}{A_1} \right)} \right]_{\mathrm{DF}} \tag{B.7}$$

and the linear time to failure dependence is, from Equation (4.25):

$$t_{\mathrm{failure}} = C \frac{A}{P_{W1} \nu_1}. \tag{B.8}$$

The temperature cycle acceleration factor was given found in Equation (4.30) as

$$\mathrm{AF_{cyclic\ fatigue}} = \left(\frac{N_2}{N_1} \right) = e^{-E_a/k_B \left(\frac{1}{T_1} - \frac{1}{T_2} \right)} \left(\frac{\Delta T_1}{\Delta T_2} \right)^K \tag{B.9}$$

and its relationship for the time to failure is similar to that for creep, as given in Equation (4.32):

$$\ln(N_{\mathrm{failure}}) = C + E_a/Tk_B - K \ln(\Delta T). \tag{B.10}$$

The acceleration factor for vibration was obtained in Equation (4.39) as

$$\mathrm{AF_{damage}} = \frac{N_1}{N_2} = \frac{T_1}{T_2} = \left(\frac{G_1}{G_2} \right)^{-b}_{\mathrm{sinusoidal}} \left(\frac{G_{\mathrm{rms}1}}{G_{\mathrm{rms}2}} \right)^{-b}_{\mathrm{random}}. \tag{B.11}$$

We note that we could substitute for the G_{rms} level the power spectral density (PSD) level, W_{psd} with an exponent $-b/2$ in Equation (4.46). The cycles to failure at any stress level gave the form of the *S–N* curves in Equation (4.42) as

$$N = CS^{-b}. \tag{B.12}$$

However, we can linearize this as

$$\ln(N_{\mathrm{failure}}) = D - b \ln(S) \tag{B.13}$$

where D is of course $\ln(C)$.

What we would like to point out is that the times to failure all have a common linearized form $Y = \mathrm{a}X + \mathrm{B}$ as

$$\ln(t_{\mathrm{failure}}) = -K \ln(\mathrm{stress}) + B \tag{B.14}$$

with an arbitrary stress acceleration factor given by

$$\mathrm{AF_{stress}} = \left(\frac{\mathrm{Stress}_1}{\mathrm{Stress}_2} \right)^{-K}. \tag{B.15}$$

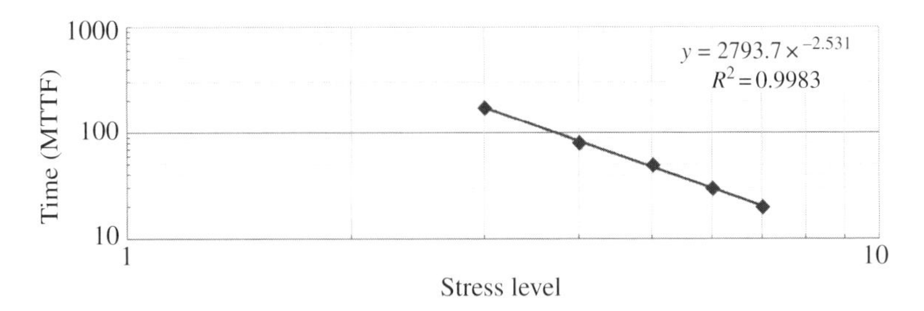

Figure B.4 MTTF stress plot of data given in Table B.2. Courtesy: DfR Software company

Table B.2 Machine stress MTTF observed

Machine stress level knob setting	Machine stress MTTF (h)
3	175
4	80
5	50
6	30
7	20

B.2.1 Example B.4: Generalized Power Law Acceleration Factors

We would like to present an example of using the power law form in Equations (B.5–B.13) by specifically using the form of Equations (B.14) and (B.15) for an unknown mechanical stress.

A machine provides an unusual stress type that is not well known. It is not covered by either of Equations (B.14) and (B.15), but we suspect as with most of the stress-related time to failure equations (except for the Arrhenius equation) that it will be governed by a power law. Table B.2 lists the failure times using a machine with knob settings of stress levels.

Assess the test results to determine if the machine's stress is causing a time to failure that has a power law form. Extrapolate the result to a knob stress level of 1.5.

Using the generalized Equation (B.14), we plot ln(MTTF) against ln(stress) as shown on the log–log plot. Using a linear regression analysis, we find the slope of the line. The result is shown in Figure B.4. Indeed, the straight line linear regression fit in the figure indicates that the power law (Equation (B.14)) is a good fit (dfrsoft, module 7, row 157 [1]).

The slope yields a K value of –2.5. We can see that it is also an inverse relationship by the negative slope. The intercept is 2793.7. Using these results we can project the MTTF to a 1.5 stress level; this will yield an MTTF of about 1000 h where $\text{MTTF}(1.5) = 2793.7\,(1.5)^{-2.5}$ (dfrsoft, module 7, row 185 [1]).

B.3 Temperature–Humidity Life Test Model

The analysis for corrosion yielded a temperature–humidity acceleration factor given in Equation (5.34) as

$$\mathrm{AF}_{TH} = \mathrm{AF}_H \, \mathrm{AF}_T. \tag{B.16}$$

The temperature acceleration factor AF_T and the humidity acceleration factor AF_H were obtained in Equation (5.35), given by

$$\mathrm{AF}_T = \exp\left\{\frac{E_\mathrm{a}}{k_\mathrm{B}}\left[\frac{1}{T_\mathrm{use}} - \frac{1}{T_\mathrm{stress}}\right]\right\}, \quad \text{and} \quad \mathrm{AF}_H = \left(\frac{\mathrm{RH}_\mathrm{stress}}{\mathrm{RH}_\mathrm{use}}\right)^M. \tag{B.17}$$

We note that the acceleration factor for humidity is interestingly enough a power law function. This expression is widely used in industry today for humidity testing to make an assessment of the accelerated time that occurs when items are put into a temperature–humidity chamber. The most common test used in the industry is a 1000 h test at 85°C and 85% relative humidity. This model in the industry is a 1986 Peck model [4]. In the Peck model from his data on corrosion testing of microelectronics in 1986, he found $M = 2.66$ and $E_\mathrm{a} = 0.7$. Often these numbers are used in the literature. The fact that the overall acceleration factor is a product of two acceleration factors indicates independence of the stresses. It also provides a fairly large acceleration factor as we will note in the next example. Although 1000 h of testing is quite common, the model suggest that only 133 h is needed to accelerate to 10 years as we will show in the next example. Note that in normal chambers, the temperature is limited to less than 100°C, the boiling point. However, highly accelerated stress test (HAST) pressure chambers are often used to allow a test temperature higher than 100°C, typically 110–130°C with humidity usually set at 85%.

B.3.1 Temperature–Humidity Bias and Local Relative Humidity

The temperature–humidity test is usually performed in microelectronics with products under minimum bias. This is because a number of failure mechanisms are driven by bias, such as silver migration, corrosion, dielectric breakdown, and so forth. That being said, minimum bias is usually tested at product specification. The minimal bias is applied so that the local relative humidity will minimally be reduced by components that generate heat. According to JESD22-A100D (see http://www.jedec.org), if the junction temperature rise of a component exceeds 10°C then it recommends that cycle bias will be a more severe test than continual bias for such parts. Usually cycle bias during humidity testing is typically performed at a rate of 12 h on and 12 h off. It is helpful to estimate the local relative humidity when calculating the acceleration factor for a particular part. This can actually be calculated using the method given in JEDEC Specification 94A (see http://www.jedec.org) or using appropriate software [1].

B.3.1.1 Example B.5: Using the Temperature–Humidity Model

Consider a temperature–humidity test at 85°C and 85% relative humidity (%RH). For a test running for 130 h, estimate the equivalent field time if the field conditions are 25°C and 40% RH using the Peck model parameters. Also assess this when a component has a junction rise of 20°C.

Part 1

Using the Arrhenius model with $E_a = 0.7\text{eV}$ and $T_{\text{test}} = 85°\text{C} = 358.15$ K, and $T_{\text{use}} = 25°\text{C} = 298.15$ K, we find

$$\text{AF}_T = 96.$$

The humidity acceleration factor is then simply 7.43. Then the overall acceleration factor is (dfrsoft, module 7, row 28 [1]):

$$\text{AF}_{TH} = \text{AF}_H \, \text{AF}_T = 7.43 \times 96 = 713.$$

For a 130 h test this equates to 713 × 130 = 10.6 years equivalency in the field.

Part 2

If the junction temperature for another part is 20°C higher, then $T_{\text{test}} = 85°\text{C} + 20°\text{C} = 105°\text{C} = 378.15$ K and $T_{\text{use}} = 25°\text{C} + 20°\text{C} = 318.15$ K. The acceleration factor is then

$$\text{AF}_T = 58.$$

To find the acceleration factor for humidity, we must now use the method JEDEC Specification 94 (see www.jedec.org). We leave the exercise mostly to the reader. However, we find that the local relative humidity on an 85% RH test at 105°C is 40% RH (dfrsoft, module 1, row 175 [1]). The same exercise can be done for the local relative humidity if the ambient temperature is 25°C and 40% RH. Then if the junction rise yields a junction temperature of 45°C, we find the local relative humidity at the die is only 13% RH. Then the humidity acceleration factor is

$$\text{AF}_H = 19.9.$$

The relative humidity acceleration factor is then (dfrsoft, module 7, row 28 [1]):

$$\text{AF}_{TH} = \text{AF}_H \, \text{AF}_T = 19.9 \times 58 = 1143.$$

For a 130 h test this equates to 1143 × 130 = 17 years equivalency in the field. This is a bit interesting. Even though the temperature acceleration factor is reduced, the humidity acceleration factor actually increased as the local relative humidity in the field is only 13% due to the junction rise for an ambient at 40% RH.

B.4 Temperature Cycle Testing

The temperature cycle given in Equation (B.9) helps to access the test time equivalency to cyclic field use conditions. The key things to be careful about when using the model are as follows in the case of solder joint issues.

- What are the numbers of cycles per day in the field?
- What is the best estimate for ΔT in the field and for the test?
- Do we need to worry about the Arrhenius factor and the frequency effect?

For solder fatigue, an alternate model was proposed by Norris and Lanzberg [5]. This model is Equation (B.9) with the frequency effect factor added for the amount of cycles per unit time observed:

$$\mathrm{AF}_{\mathrm{TC\ cyclic\ fatigue}} = e^{-E_a/k_B\left(\frac{1}{T_1}-\frac{1}{T_2}\right)}\left(\frac{\Delta T_{\mathrm{stress}}}{\Delta T_{\mathrm{use}}}\right)^{K}\left(\frac{F_{\mathrm{use}}}{F_{\mathrm{stress}}}\right)^{K}. \tag{B.18}$$

The use of this model for the frequency effect is as follows. For lead-free SAC (Sn/Ag/Cu) solder, if the maximum upper cycle temperature is greater than 1/2 of the melting temperature (*c*. 216°C/2 = 108°C) and if the cyclic frequency is longer that about 17 min, the frequency effect is said to come into play. Below this temperature there is little effect on fatigue (we ignore the frequency effect) as creep plays less of a role.

For non-solder fatigue (where the frequency effect is not used), most users ignore the temperature Arrhenius factor which we found should be added in Chapter 4.

B.4.1 Example B.6: Using the Temperature Cycle Model

A temperature-cycle test is performed for 200 cycles from –25°C to +125°C. How many cycles is this in the field at 1 or 2 cycles a day for delta use of 0–35°C? Assume a conservative temperature cycle exponent of 2.0.

Analysis: The temperature cycle ΔT factor is (dfrsoft, module 7, row 42 [1])

$$\mathrm{AF}_{\mathrm{TC}} = \left(\frac{\Delta T_{\mathrm{stress}}}{\Delta T_{\mathrm{use}}}\right)^{K} = \left(\frac{150}{35}\right)^{2} = 18.4.$$

For the Arrhenius part we use the upper temperature for field and use condition of 35°C and 125°C. The activation energy commonly used for solder fatigue is 0.123 eV. We have not really specified what we are actually testing. Often we may not know the actual Arrhenius activation energy. If we are concerned with creep and fatigue as in Equation (B.18), then perhaps this would be a good number. Using it we find the Arrhenius part is 2.85 (dfrsoft, module 7, row 81 [1]).

Then we can compare with and without the Arrhenius factor. Without the Arrhenius factor the field years = 18.4 × 200/365 = 10 years and with it the field years = 28.5 years.

Now if the product was exposed to 2 cycles per day in the field, this would be equivalent to 14.25 years.

B.5 Vibration Acceleration

There are two main types of vibration tests where one simulates field use conditions; these are sine or random vibration.

Sine vibration testing has products exposed to sinusoidal vibrations as for type shown in Figure B.5. Sine vibration is usually specified in terms of G where the unit G denotes acceleration of gravity (9.8 m/s^2). If for example a system is in free fall, then this is 1 G. In sine vibration testing, products are mounted on an electrodynamic shaker and often certain specifications pertain. For example, MIL-STD 810E is one of the most common specifications. There are two

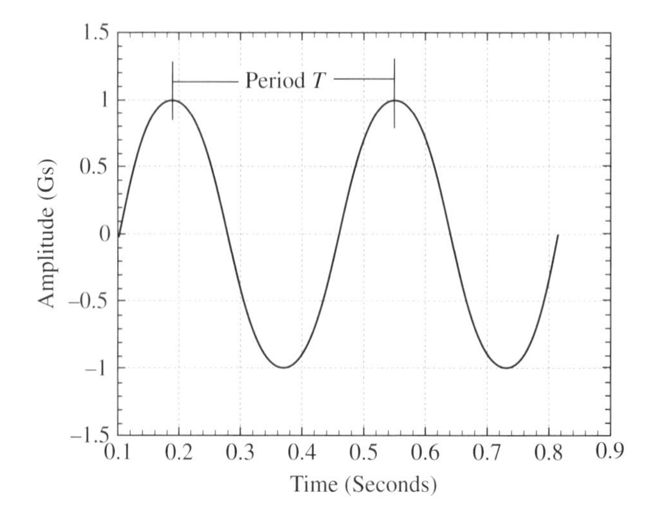

Figure B.5 Sine vibration amplitude over time example

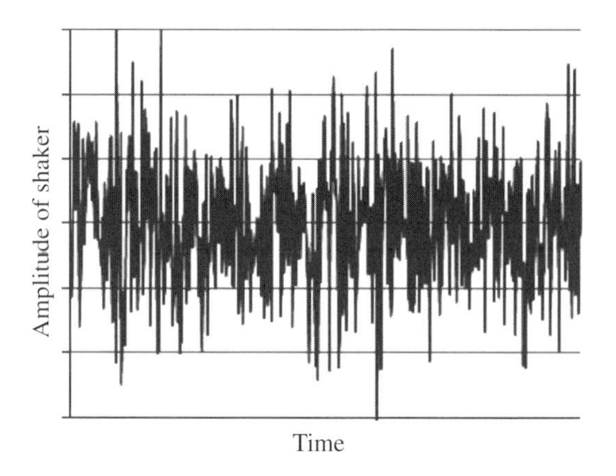

Figure B.6 Random vibration amplitude time series example

main types of sine tests: these are a sine sweep and the other is a resonance dwell. Often one use sine sweep to look for resonances.

Typically, the motivation for doing sine testing is to simulate field use conditions that are sinusoidal-like. Highly periodic sine-like vibrations occur for say electric motors, helicopter periodic blade motion, or other similar propeller-type aircraft. Some vibration types such as helicopters are mixed and require both sine and random vibration testing. This is often termed sine on random (dfrsoft, module 5, row 24 [1]).

Random vibration testing is usually performed to simulate field use condition. As the name suggests, the vibration is truly random. This is displayed in Figure B.6. Unlike sine vibration

Table B.3 Gaussian probabilities (%)

1σ	2σ	3σ	4σ
Two-sided probability			
68.27	95.45	99.73	99.99
Percent above			
31.73	4.55	0.27	0.006

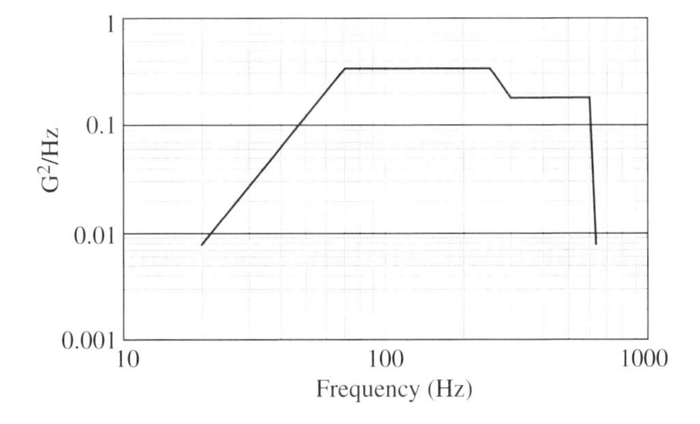

Figure B.7 PSD of the random vibration time series in Figure B.6

where the amplitude is known with a specific G-level, random amplitude happens with a probability of occurrence. A common type of random vibration often observed in the field is Gaussian. Table B.3 displays the probabilities for Gaussian vibration. For example, the amplitudes will be above 3σ about 0.27% of the time and above 4σ just 0.006% of the time. Note Gaussian vibration has a mean of zero so σ (which is the G_{rms} content for Gaussian vibration) is the actual amplitude in terms of G_{rms}. For a random vibration of 60 s the amount of the time at which the amplitude will be greater than 1σ is 60 s × 0.3173 = 18.22 s; the amount of time at which the amplitude will be less than 1σ is 60 s × 0.683 = 41 s.

Another aspect of random vibration is the concept of going from the time series spectrum in Figure B.6 and performing a Fourier transform to obtain the frequency spectrum graph shown in Figure B.7 (dfrsoft, module 25, row 71 [1]). The frequency spectrum plot is termed the PSD plot. This graph looks a lot easier to interpret than the time domain plot; test specifications are often provided in the frequency domain for this reason. The area under the PSD curve is the G_{rms} content. When doing accelerated testing one can use either the PSD peak levels or the G_{rms} content. The equations are almost the same. For the interested reader this is the difference between using Equations (4.39) and (4.45), where the exponent is divided by two resulting from finding the area under the PSD curve. For example, the value of $b \approx 5$ is commonly used for electronic boards. However, a conservative value for the fatigue parameter b is about 8 (e.g., $M_b = 4$). MIL STD 810E (method 514.4–46) recommends $b = 8$ for random loading.

Random vibration is also performed on an electrodynamic shaker which is capable of accelerated testing on products requiring either sine or random vibration. Often specifications from MIL STD 810 are used. Alternately, one can design their own test to simulate field conditions.

B.5.1 *Example B.7: Accelerated Testing Using Sine and Random Vibration*

An accelerated test is required to simulate 10 years of life for sine testing at resonance occurring at 100 Hz. The accelerated test is specified for 4 *G* of sine vibration where in the field the product is only exposed to a 1.5 *G* level at a resonance of about 2% of the time. Use a sine vibration exponent of 5.

Estimate the time to simulate 10 years of life for an assembly that is to be tested at a PSD level of 0.12 G^2/Hz from 0 to 2000 Hz. It is estimated that the assembly will undergo a worst case of 1% exposure of 0.03 G^2/Hz over this bandwidth. Use MIL STD 810 $M_b = 8$.

Sine Vibration Analysis
The sine vibration acceleration factor is that of Equation (4.39), also given in Section B.2.4 is

$$\mathrm{AF}_{\mathrm{damage}} = \left(\frac{G_1}{G_2}\right)^{-b}_{\mathrm{sinusoidal}} = \left(\frac{1.5}{4}\right)^{-5} = 335.5.$$

The test for 10 years in the field is about 10 × 8760 × 2%/335.5 = 5.22 h (dfrsoft, module 25, row 158, col B [1]).

Random Vibration Analysis
The random vibration acceleration factor is from Equation (4.45):

$$\mathrm{AF}_{\mathrm{damage}} = \left(\frac{\mathrm{PSD}_{\mathrm{use}}}{\mathrm{PSD}_{\mathrm{stress}}}\right)^{-b/2}_{\mathrm{random}} = \left(\frac{0.03}{0.12}\right)^{-4} = 256.$$

The test for 10 years in the field is about 10 × 8760 × 1%/256 = 3.4 h (dfrsoft, module 25, row 158, col C [1]).

B.6 Multiple-Stress Accelerated Test Plans for Demonstrating Reliability

Multiple test demonstration testing of the type shown in Figure B.2 is set up to demonstrate an overall failure rate with a certain level of confidence that helps determine the sample size for each test. The overall goal is to demonstrate a certain failure rate over multiple stresses that the product might be subjected to in the field. The most common stress that are tested for microelectronics is high-temperature operating bias (see Section B.1), humidity–temperature bias (see Section B.3), and temperature cycle (see Section B.4). Further, if we are concerned with vibration (Section B.5) than this must be included as well. One of the key obstacles in designing such a test is to allocate the failure rate among the tests. For example, if the desired overall failure rate is say 0.6 FMH, than one possible scenario is to allocate a portion of the failure rate evenly among the tests; in this case that would be 0.2 FMH. Another possible scenario is to optimize the sample size (see example below) among the three tests given constraints on the failure rate and the total sample size. The following example illustrates these methods.

B.6.1 Example B.8: Designing Multi-Accelerated Tests Plans: Failure-Free

An example of equal-failure-rate test allocation is provided in Table B.4. The three stress environments that a product is exposed to are accelerated to demonstrate a failure rate of 0.6 FMH. Using the method in Section B.1.4, one can verify these numbers. The reader may wish to check the values in the table as an exercise (dfrsoft, module 8, row 28, col C, and row 41, col. B [1]). The test constraints are as follows.

- Total failure allowed = 0;
- Confidence level = 90%;
- Failure rate objective = 0.6 FMH;
- High-temperature operating life (HTOL): 1000 h, $T_{use} = 50°C$, $T_{stress} = 125°C$, $E_a = 0.7$ eV;
- Temperature cycle (TC): 200 cycles, $\Delta T_{use} = 30°C$, $\Delta T_{stress} = 150°C$, exponent 2;
- Temperature–humidity bias (THB): 1000 h, $RH_{use} = 40\%$, $RH_{stress} = 85\%$, $T_{use} = 35°C$, $T_{stress} = 85°C$.

Table B.5 below provides another example where equal allocations are not used but we instead seek to optimize for a minimum sample size among the three tests given the same test conditions and constraint of 0.6 FMH objective. As an exercise, the reader may wish to check

Table B.4 Multiple stress accelerated test to demonstrate 0.6 FMH

Stress	Chapter example number	Acceleration factor	Test time	FMH	Minimum sample size
HTOL	B.1	114	1000 h	0.2	79
THB	B.5	295	1000 h	0.2	39
TC	B.6	25	200 cycles	0.2	96
Total				0.6	236

Courtesy: DfR Software company.

Table B.5 Optimized multiple stress accelerated test for 0.6 FMH

Stress	Example number	Acceleration factor	Test time	FMH	Minimum sample size
HTOL	B.1	114	1000 h	0.233	87
THB	B.5	295	1000 h	0.144	54
TC	B.6	25	200 cycles	0.226	85
Total				0.6	226

Courtesy: DfR Software company.

the optimized results in the table (dfrsoft, module 8, row 28, col C, and row 39, col. L [1]). Comparing Tables B.4 and B.5 we see the optimized results saved about 10 devices.

- Total failure allowed = 0;
- Confidence level = 90%;
- Estimate life each test greater than 10 years;
- Failure rate objective = 0.6 FMH;
- HTOL 1000 h, $T_{use} = 50°C$, $T_{stress} = 125°C$, $E_a = 0.7$ eV;
- TC 200 cycles, $\Delta T_{use} = 30°C$, $\Delta T_{stress} = 150°C$, exponent 2;
- THB: 1000 h, $RH_{use} = 40\%$, $RH_{stress} = 85\%$, $T_{use} = 35°C$, $T_{stress} = 85°C$.

B.7 Cumulative Accelerated Stress Test (CAST) Goals and Equations Usage in Environmental Profiling

In Section 4.4 we described cumulative accelerated stress test (CAST) goals, listed in Table 4.3. To illustrate this for accelerated testing, one needs to estimate the field use conditions. Often, stresses in the field vary over time. For example, a product during a year in the field might be temperature-profiled over say 1 year as indicated in Table B.5.

When this occurs, what stress temperature do we use for an accelerated test and how do we plan it? To this end we now provide an example.

B.7.1 Example B.9: Cumulative Accelerated Stress Test (CAST) Goals and Equation in Environmental Profiling

Consider the 1 year profile in Table B.5 worked out prior to an accelerated test. The reliability engineer needs to figure out a CAST goal for the accelerated test equivalent to infant mortality of 1 year and a life test goal for 10 years. He now uses the temperature CAST goal equation in Table 4.3 that allows him to do so, which is

$$\tau_1 = \sum_{i=1}^{k} \frac{t_i}{\mathrm{AF}_{\mathrm{damage}}(1,i)}.$$

First he defines τ_1 as 1 year infant mortality time. Then he starts by estimating a reference stress, let's try 50°C for AF(1, i). Next he uses the Arrhenius acceleration factor with an activation energy of 0.7 eV for the above equation. With this he simply plugs in the values in Table B.5 to come up with the actual results for τ_1 which is equivalent to 8474 h at 50°C. The reader should verify this time (dfrsoft, module 9, row 8, col A, [1]). He can then iterate and refine the reference stress; the actual temperature in this case is very close and is 49.57°C for exactly 1 year.

The results of the above equation are shown in Table B.6. For example, dividing the acceleration factor AF (30°C, 49.57°C) = 5.08, shown in column 4, into 2190 gives 431 hours equivalent, as show in column 5. That is, if our environment is profiled at 49.57°C and we did a test at 30°C for 2190 hours, it would equate to 431 hours at 49.57°C. We continue this

Table B.6 Profile of a product's temperature exposure per year

Temperature	Time/year	Percent of time (%)	AF to 49.57°C	Profiled time at 49.57°C
30	2190	25	5.08	431
40	2628	30	2.16	1219
50	2190	25	0.97	2258
60	1314	15	0.46	2891
70	438	5	0.22	1962
Sum	8760	100		8760

and note that the sum in column 5 is indeed 1 year equivalent. That is, only this unique profiled temperature equates to 1 year. The reader should verify this temperature. We wanted to simplify the explanation so we used this iterative method.

Next the engineer wants to design a 10 year life test. We can simply multiply up the profile in Table B.5 to a 10 year CAST goal and the results will yield the same reference temperature requirement of 49.57°C. With this CAST goal, we can estimate the device hours needed in an accelerated 10 year life test at any reasonable test temperature to achieve enough device hours to verify the 10 year goal relative to the reference temperature of 49.57°C. For example, using the chi-squared method in Section A.8, we have for an accelerated test at 120°C

$$\bar{\lambda} = \frac{\chi^2\{\gamma; 2y+2\}}{2n\,\mathrm{AF}(49.57^\circ\mathrm{C}, 120^\circ\mathrm{C})\,t}.$$

This is then subject to the restriction that AF × test time = 10 years. We now can trade off between the sample size n and the failure rate λ objective we might need given a certain level of confidence γ and allowed failures y (usually 0 or 1; see Table B.6).

References

[1] Author's software, download at http://www.dfrsoft.com (see Section A.1.1 for details) to follow along with examples.

[2] O'Connor, P. and Kleyner, A. (2012) *Practical Reliability Engineering*, 5th edn, John Wiley & Sons, Ltd, Chichester.

[3] Feinberg, A., Crow, D. (eds) *Design for Reliability*, M/A-COM 2000, CRC Press, Boca Raton, 2001.

[4] Peck, D.S. Comprehensive model for humidity testing correlation. In *Proceedings of the 24th IEEE International Reliability Physics Symposium*, April 1986, Anaheim, CA, pp. 44–50.

[5] Norris, K.C. and Landzberg, A.H. (1969) Reliability of controlled collapse interconnections. *IBM Journal of Research and Development*, **13** (3), 266–71.

Special Topics C

Negative Entropy and the Perfect Human Engine

C.1 Spontaneous Negative Entropy: Growth and Repair

While devices and systems that we use every day will not spontaneously repair themselves, Mother Nature has apparently provided life forms with this capability. In fact, we can think of growth as a sort of negative entropy which occurs in the first part of our lives. The overall growth or repair process must still generate positive entropy. Still, in our definition of entropy for a system, we understand spontaneous positive entropy change $\Delta S > 0$ as the amount of disorganization that occurs; then spontaneous negative entropy change

$$\Delta S_{\mathrm{N}} < 0 \tag{C.1}$$

is a term that we might argue can apply to Mother Nature's life forms. In human life, as we grow we become more organized; we have a larger capability for doing more useful work so our free energy is essentially increasing. In a sense we can label this as spontaneous negative entropy. Furthermore, when we are injured, our bodies will try and repair the damage by creating a spontaneous amount of negative entropy equal to or greater than the entropy damage (see Equation (1.13)) change that occurred during our injury. This quantity, in thermodynamic terms, is estimated in Equation (1.14).

Negative entropy was first introduced by Erwin Schrödinger in a non-technical field in his 1944 popular-science book *What is Life* [1]. Schrödinger uses it to identify the propensity of the living system to want to organize, which is contrary to the second law. That is, for most of us, we like to build houses, build cities, and organize our way of life. This is also observed in lower life forms. So this book's discussion on the subject was limited to such areas. The subject dis-

Thermodynamic Degradation Science: Physics of Failure, Accelerated Testing, Fatigue, and Reliability Applications, First Edition. Alec Feinberg.

cussed here is not found easily in the literature and is an atypical thermodynamic term as we have described it.

It is not immediately obvious why spontaneous negative entropy processes would be allowed by the second law of thermodynamics.

The spontaneous tendency of a system to go toward thermodynamic equilibrium cannot be reversed without at the same time changing some organized energy, work, into disorganized energy, heat.

Some would argue that repair and growth are not spontaneous negative entropy. Certainly repair does not reverse time; repair is done by removing the damaged area and re-grows the cells as close to their original growth state as nature permits. However, growth and repair are not easily controlled either which makes them more or less spontaneous events. So, from this perspective at least, we can think of it as spontaneous negative entropy. The key phrasing of the second law states that there is a natural tendency "to come to thermodynamic equilibrium." Mother Nature likely creates a closed system that encourages growth and repair in order for the system to come to some sort of final growth or repair equilibrium condition. This means that creating a closed system takes energy, that is, part of the disorganized entropy production. In the non-equilibrium state we can think of the organized growth or repair as

$$\frac{d\Delta S_{\text{system-growth or repair}}}{dt} < 0, \qquad \frac{d\phi_{\text{system-growth or repair}}}{dt} > 0. \tag{C.2}$$

Growth or repair then stops when the system is in thermodynamic equilibrium or perhaps a quasi-equilibrium state. Growth and repair must obey the second law. Therefore, the overall entropy production is positive, yet the results are organized matter.

But what is being disorganized then that occurred in the repair or growth process? The growth and repair processes are similar to that discussed in Section 1.9.1. Apparently, it can be thought of like the repair of a device: the system undergoes a seemingly negative entropy change, but with processes that must have a tendency to come to thermodynamic equilibrium. Therefore entropy is being maximized in growth but with the results of organized matter; overall, the universe has a positive entropy change equal to or greater than the negative entropy change. Energy has been used to create and repair the living system. Mother Nature has converted fuel (food) which becomes decomposed to build or repair our bodies. The process is similar to repair of a device, in that the body is manufacturing cells for growth and repair. The unused decomposed mater is waste that adds entropy to the universe. A propose model is provided here.

C.2 The Perfect Human Engine: How to Live Longer

In this section we would like to make use of some of our knowledge that we have built up in thermodynamic degradation and see if we can apply it to the miracle of the human engine. Although this is not the perfect science for such a task, as Mother Nature is very complex, we can use the opportunity to exemplify some strong analogies that exist as well as some major differences. To this end, we can gain insight that can help us manage our human engine and perhaps live a little longer.

If we were to invent the perfect human engine, what characteristics would it have? Our engine objectives might be: high work efficiency; long life; and the ability to repair damage close to near-perfect original conditions.

C.2.1 Differences and Similarities of the Human Engine to Other Systems

To understand how we might make such improvements, one might first ask how a human engine is different from say a heat engine. At first glance there are certainly some similarities.

- They have similar lifecycles and follow the bathtub curve: this gives us some indication that there are indeed similarities to engines and electronic devices.
- They both have cyclic characteristics. When considering the digestive cycle, the fuel cycle immediately comes to mind, that is, taking in fuel and depositing waste and repeating. Both can be thought of as having cyclic efficiencies, taking in fuel and putting out a certain amount of work. Fuel choice can help the efficiency of the engine or hurt it. Body work cycle: The human engine's work can be viewed as cyclic (daily work, daily rest, and repeat). Brain cycle: Performing daily mind work, resting and repeating. Heart cycle: typically 50–100 beats/min. Breathing cycle: inhale–exhale average of 12 breaths/min.
- They both can sustain cyclic and non-cyclic damage, increasing disorder.
- They both have a finite lifetime for which they can perform cyclic work.

As with many life forms created by Mother Nature, the humane engine clearly differs from engines and devices in many key aspects; we will discuss two significant differences here: (1) growth: a human engine takes in fuel and uses it for growth during the first part of its life; and (2) self-repair: when damage occurs to its engine, especially minor damage, it can repair itself to various degrees of precision.

C.2.2 Knowledge of Cyclic Work to Improve Our Chances of a Longer Life

Some of the thermodynamic analogies at this point may seem obvious to improve human engine life, but they are worth noting. Clearly, irreversible damage reduces efficiency. As noted in Equation (2.88):

$$\eta = \frac{W_{\text{actual}}}{W_{\text{actual}} + W_{\text{irreversible}}}. \tag{C.3}$$

Damage is an enemy as it creates irreversibility and reduces our ability to change fuel into work. Although some damage is reversible due to repair, repair inefficiencies also occur and irreversibilities build up over time creating aging. Inefficient fuels are also poor choices, creating inability to convert fuel into work and causing excess waste.

In terms of cumulative damage, we have noted that stress is a key factor in increasing work cycles to failure. Some examples of stress for the human engine cycles are as follows.

Table C.1 Human cyclic engine and possible stresses that shorten cycle life

Human engine cyclic subsystem	Local interacting environment	Work cycle	High-amplitude stress cycles
Digestive system	Food	24-hour daily cycle of chewing, energy conversion digestion, waste	Hard to digest foods, large size foods, overeating
Body work cycle	External work	24-hour daily workloads: rest, repeat	High stress activity interruption, abrupt change to daily work
Heart cycle	Blood, air	50–100 beats/min, expansion–contraction work to circulate blood	High stress activity and lack of oxygen contribute to a shorter life; moderate exercise is better than heavy exercise
Breathing: lungs cycle	Air	About 12 breaths/min, inhale–exhale	Air quality excessive or unusual exhaustion
Brain cycle	Five senses	24-hour daily cycle, awake–sleep–other	Change in daily cycle of mind activity, high stress on one or more of the five senses, excessive work load, high emotional stress
Repair cycle	Body, blood, oxygen, nutrients	Damage-repair, daily regeneration	Poor blood flow and lack of oxygen and nutrients creates additional stress, excess damage due to trauma injury

- Digestive cycle: Environment food-stress might include hard-to-digest foods (processed foods); large-sized food not initially reduced in size (suggesting pureed food may be helpful); high-stress cycles such as eating large quantities of food at one time (excessive overeating).
- Body work cycle: high-stress cycles include activity that can cause damage (football, excessive exercise, extreme sports); and cyclic interruption, abrupt change to daily work, daily rest and repeat.
- Heart cycle: high-stress cycle for long periods such as marathon training compared to light joggers [2].
- Breathing cycle: air quality can cause stress; excessive athletic use cycles (e.g., marathon running).
- Brain cycle: change in daily cycle of mind work, resting and repeating; excessive work load; high emotional stress.
- Repair cycle: damage–repair daily regeneration cycle; excess damage loads and lack of nutrients, oxygen, and blood flow can increase repair cyclic stress.

These ideas are summarized in Table C.1. Some ideas may be counterintuitive such as high-stress exercising. One study on the human heart suggested a U-shaped stress versus life cycle curve that we discuss in the following section. What we know from cyclic work and Miner's rule can seem at times related to human life cycle; reducing stress loads can increase longevity; inactivity also reduces life span, however. Thus, a U-shaped *S–N* curve seems appropriate. In other areas there may be more parallels to engines. For example, there have been suggestions that eating less but more frequently would be beneficial. In cyclic theory this makes sense. Table C.1 provides a cyclic summary.

C.2.3 *Example C.1: Exercise and the Human Heart Life Cycle*

The reader may be skeptical of the results related to high stress activity on the heart, as exercise and lifespan are correlated. However, the following is the result of a Copenhagen City Heart Study by Schnohr *et al.* [2]. A total of 1098 healthy joggers and 3950 healthy non-joggers were followed from 2001 with the study published in February 2015.

> "[Results:] Compared with sedentary non joggers, 1 to 2.4 h of jogging per week was associated with the lowest mortality (multivariable hazard ratio [HR]: 0.29; 95% confidence interval [CI]: 0.11–0.80). The optimal frequency of jogging was 2 to 3 times per week (HR: 0.32; 95% CI: 0.15–0.69) or less than or equal to 1 time per week (HR: 0.29; 95% CI: 0.12–0.72). The optimal pace was slow (HR: 0.51; 95% CI: 0.24–1.10) or average (HR: 0.38; 95% CI: 0.22–0.66). The joggers were divided into light, moderate, and strenuous joggers. The lowest HR for mortality was found in light joggers (HR: 0.22; 95% CI: 0.10–0.47), followed by moderate joggers (HR: 0.66; 95% CI: 0.32–1.38) and strenuous joggers (HR: 1.97; 95% CI: 0.48–8.14).
>
> "[Conclusions:] The findings suggest a U-shaped association between all-cause mortality and dose of jogging as calibrated by pace, quantity, and frequency of jogging. Light and moderate joggers have lower mortality than sedentary non joggers, whereas strenuous joggers have a mortality rate not statistically different from that of the sedentary group."

At least in this one study, verification seems to indicate that Miner's type rule for human mortality very roughly has some validity. Obviously, it is clear that low stress will not lead to longer life as in the case of metal fatigue. The study suggested a U-shaped model. We have taken their U-shaped description as an initial idealized *S–N* curve model here (not given in their study) of the type plotted in Figure C.1 where light to moderate stress increases life cycle expectancy for the human heart. The U-shaped curve is just meant to illustrate the major difference between human fatigue and fatigue in metals. It is reasonable to assume that a major difference between metal fatigue damage versus human damage is the repair factor, which is discussed in the next section. If we can completely repair our daily damage, then we could live forever. On the other hand, if we could not repair ourselves, then it is likely that human heart fatigue would follow closely to the metal fatigue line. We know that the ability to do repair is also related to exercise; without it, the body cannot properly do its repair job. Yet too much exercise will cause excess damage that the repair cycle cannot keep up with (Figure C.1).

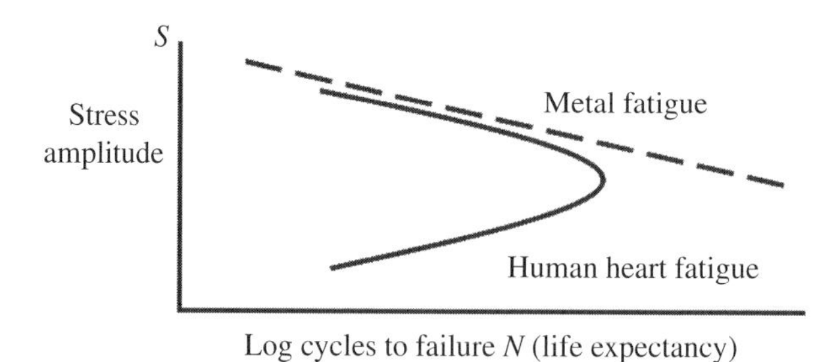

Figure C.1 *S–N* curve for human heart compared to metal *N* fatigue cycle life versus *S* stress amplitude

C.3 Growth and Self-Repair Part of the Human Engine

Growth and self-repair indicate significant differences between the human engine and a normal engine or device. First of all, growth is unlike what is expected from the second law of thermodynamics. We can say that the human engine in the early part of life is becoming more ordered, and its free energy is actually increasing in the sense that the amount of useful work that the human engine can perform increases as we grow older. We then have a case that for growth the human engine is becoming more ordered during the growth phase, and entropy must be decreasing:

$$\Delta S_{\text{human engine}} < 0, \quad \text{for } 0 < \text{time} < \text{human growth phase.} \tag{C.4}$$

Furthermore, when damage occurs, disorder is created in the human system; repair then reverses damage with a certain amount of efficiency. However, prior to repair, damage increased entropy:

$$\Delta S_{\text{human damage}} > 0, \quad \text{for } 0 < \text{time} < \text{to repair starts.} \tag{C.5}$$

Then when repair starts, damage becomes somewhat reversible:

$$\Delta S_{\text{human repair}} < 0, \quad \text{for repair start} < \text{time} < \text{repair completed.} \tag{C.6}$$

The exchange of entropies in repair is

$$\Delta S_{\text{gen}} = \Delta S_{\text{repair}} + \Delta S_{\text{env}}, \quad \Delta S_{\text{gen}} > 0. \tag{C.7}$$

The total entropy of any process increases; in keeping with the second law (Equation (1.4)) $\Delta S_{\text{gen}} \geq 0$. Since $\Delta S_{\text{repair}} < 0$, than the entropy damage to the environment must be positive and greater than the negative repair entropy by the second law, that is:

$$\Delta S_{\text{env}} \geq \Delta S_{\text{repair}}. \tag{C.8}$$

(Also see Section 1.9.1.) Since the repair process is in itself cyclic, this means that the internal energy needs to be restored to its original state. Its change due to the damage portion of the cycle is:

$$\Delta U_{\text{change-due-to-damage}} = \int_0^{\text{damage}} dU = U_{\text{damage}} - U_{\text{non-damage}}. \tag{C.9}$$

We can make a repair model. In a simplified view, the body is essentially a battery that gets charged in the repair process by the amount above. Since it must discharge nutrients adding heat and work to repair, the internal energy change is essentially in a damage–repair cycle by the combined first and second law (see Equation (2.60)):

$$\oint dU = \oint \delta W + T\oint dS + \Delta U_{\text{unrepaired}} = 0. \tag{C.10}$$

That is, for a perfect repair the internal energy in the cycle is unchanged; an imperfect repair leaves us with some inefficiency and permanent change to the internal energy, however. Daily use causes damage, which also requires repair. The unrepaired portion builds up, causing body fatigue and damage.

C.3.1 Example C.2: Work for Human Repair

We can make a simplified thermodynamic repair model for instructional purposes. The repair process is shown in Figure C.2. An injury occurs, and after a few hours the entropy is at a maximum where the entropy damage is S_{damage} and the area is infected with temperature rise T_{high}. Repair work is done W_{repair} and the injury is almost completely repaired; the unrepaired entropy is $S_{\text{unrepaired}}$. The change to the internal energy from repair cycle ΔU is due to the unrepaired damage. In this case from the first and second law the minimum repair work is:

$$W_{\text{repair-min}} = T_{\text{high}} S_{\text{damage}} - \Delta U_{\text{damage-unrepaired}} = T_{\text{high}} S_{\text{damage}} - T_{\text{low}} S_{\text{unrepaired}}. \tag{C.11}$$

The negative entropy generated for the repair process is

$$S_{\text{gen}} = S_{\text{damage}} - S_{\text{unrepaired}} = -S_{\text{repair}} \tag{C.12}$$

where S_{repair} is the negative entropy needed for the repair process. That is, in the case of repair:

$$S_{\text{damage}} - S_{\text{unrepair}} + (-S_{\text{repair}}) = 0 \quad \text{and} \quad S_{\text{unrepaired}} = S_{\text{damage}} + (-S_{\text{repair}}) \tag{C.13}$$

so that the minimum work in the repair process is found by combining these equations

$$W_{\text{repair-min}} = S_{\text{damage}} (T_{\text{high}} - T_{\text{low}}) - T_{\text{low}} (-S_{\text{repair}}). \tag{C.14}$$

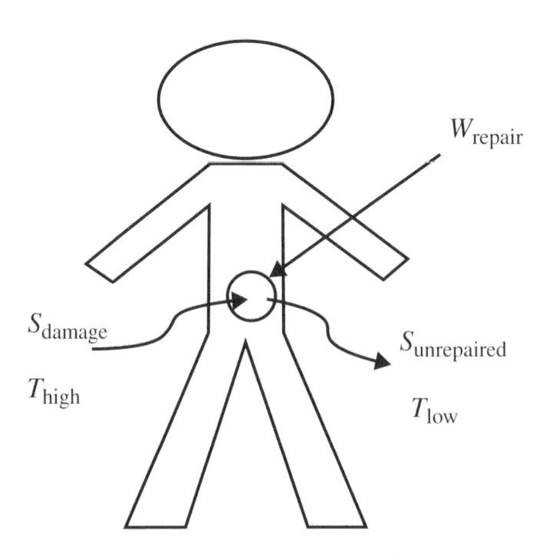

Figure C.2 Simplified body repair

In the case of perfect repair $S_{\text{damage}} = -S_{\text{repair}}$ and the minimum repair work is

$$W_{\text{repair-min}} = S_{\text{damage}} T_{\text{high}} = Q_{\text{heat}} \tag{C.15}$$

where Q_{heat} is the heat dissipated prior to repair which can be measured.

The efficiency of repair is

$$\eta_H = \frac{W_{\text{actual}}}{W_{\text{actual}} + W_{\text{irreversible}}} = \frac{W_{\text{repair-min}}}{W_{\text{repair-min}} + T_{\text{low}} S_{\text{unrepaired}}} = 1 - \frac{T_{\text{low}}}{T_{\text{high}}}\left(1 + \frac{S_{\text{repair}}}{S_{\text{damage}}}\right). \tag{C.16}$$

In the case of perfect repair where $S_{\text{repair}} = -S_{\text{damage}}$, the efficiency is 1. Since $S_{\text{repair}} = -fS_{\text{damage}}$, a fraction f of S_{damage}, then their ratio ($S_{\text{repair}}/S_{\text{damage}}$) will give a value between 0 and 1, and this obeys the relation in Equation (3.15) where

$$\eta \leq 1 - \left(\frac{T_{\text{min}}}{T_{\text{max}}}\right),$$

similar to a cyclic heat engine process.

C.4 Act of Spontaneous Negative Entropy

When we repair or manufacture a device, the device becomes more organized so the entropy of the device has decreased. But the process to build or repair caused excess entropy to the universe. In the same way, energy is used to create and repair a living system. Mother Nature uses some fuel (food) for energy needed for growth and/or repair, and the food decomposes into waste which is an increase in entropy production. The decomposed matter is waste that adds entropy to the universe. This helps to justify spontaneous negative entropy as not violating of the second law. The act of spontaneous negative entropy can be clarified somewhat as follows.

> *The act of negative spontaneous entropy found in living systems is the result of initial work done by the living system to create a tendency for the living system to come to equilibrium with its neighboring living environment that drives the living system to a state of repair or growth in order to achieve thermodynamic equilibrium. The overall process of entropy production generated between the environment and the system is still positive, in keeping with the second law.*

If we concur with this statement, where we are trying to maintain agreement with the second law to have a tendency to come to thermodynamic equilibrium, it means that our human engine must create an environment that encourages spontaneous negative entropy. How does this come about? We can hypothesize that our body takes in food and converts it to work, charging up an area that drives the need for repair or growth. In this sense there is not much difference between repair and growth. We know that an injured area will not reverse itself and become organized, seemingly going back in time; it must be removed and re-grown. This is similar to device repair.

The key phrasing of the second law is then: there is a natural tendency "to come to thermodynamic equilibrium." Then we can surmise that Mother Nature creates a closed system that encourages growth and repair in order for the system to come to some sort of final equilibrium condition of full growth or full repair. In the non-equilibrium state:

$$\frac{dS_{\text{growth or repair}}}{dt} < 0.$$

Growth or repair then stops when the system is in thermodynamic equilibrium or perhaps a quasi-equilibrium state.

In order to improve our understanding and remain within the scope of this book, we can hypothesize part of the equilibrium process from our prior models in Chapters 1 and 8. Once the nutrients are brought to the repair site, they likely diffuse into the areas under diffusion forces. Diffusion can be driven both by a concentration gradient and an electrical charge across the repair area [3]. This is one possible scenario, that is, we can hypothesize something like the equation below where we have combined Chapter 2 (Equation (2.75)) and Chapter 8 (Equation (8.8)) results to obtain

$$dS_{\text{total}} = \left(\frac{1}{T} - \frac{1}{T_{\text{env}}}\right)dU + \left(\frac{E}{T_{\text{env}}} - \frac{V}{T}\right)dq + \left(\frac{\mu_{\text{env}}}{T_{\text{env}}} - \frac{\mu}{T}\right)dn = -dS_{\text{repair}}. \tag{C.17}$$

Here energy flow will go from the higher temperature area $T_{\text{env}} > T$ so that the repair energy is increasing $dU > 0$. If $E > V$ then $dq > 0$ and the repair area is losing charge; for $\mu_{\text{env}} > \mu$, $dn > 0$ the repair area is receiving particles. When $T = T_{\text{env}}$, $E = V$, and $\mu_{\text{env}} = \mu$, the repair process is completed and we are in thermodynamic equilibrium. The result is a more organized area. The environment in a sense appeared to increase entropy, yet the repair or growth area became more organized. In this model, the body is tricked into increasing entropy change dS_{total}, that caused more organization dS_{repair} than disorganization in order to come to equilibrium with the neighboring environment.

This is just meant to demonstrate how a spontaneous repair or growth could be taking place in a living system in agreement with principles of the second law. There are so many unanswered questions, such as how does the body know what cells match for repair and how does the body set up a spontaneous tendency for repair and growth?

C.4.1 Repair Aging Rate: An RC Electrical Model

We might assert that the rate of negative repair entropy varies as the entropy damage to within an aging factor $f(t)$:

$$\frac{dS_{\text{repair}}}{dt} = f(t)S_{\text{damage}}. \tag{C.18}$$

For increasing aging, we propose that $f(t)$ increases where $0 < f(t) < 1$. Therefore, the rate of change of $f(t)$ is some function of the unrepaired entropy damage $S_{\text{unrepaired}}(t)$ that builds up over our lifetime and reduces our ability to heal at an older age. As we grow older, our ability to heal completely is reduced and $f(t)$ changes. If this was not the case and we had complete repair every day, we would not age. As $f(t)$ decreases, then according to this proposed equation

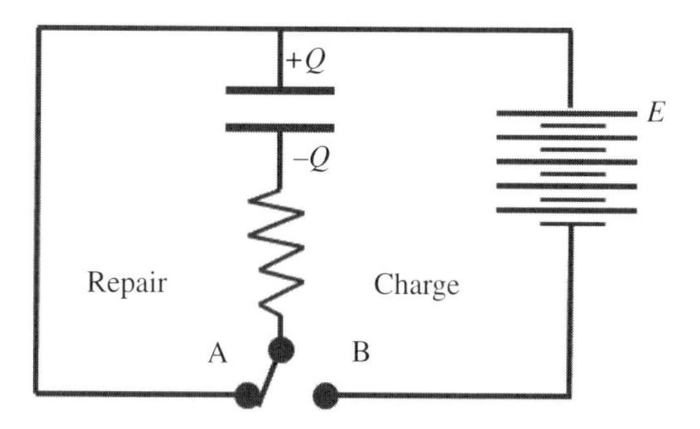

Figure C.3 Charge and repair RC model for the human body

the repair rate will also slow. The model is simple but illustrates a possible view of aging. From our experience, $f(t)$ must be a slow function of time compared to the repair rate, that is,

$$\frac{d\Delta S_{\text{repair}}}{dt} >> \frac{df(t)}{dt}. \tag{C.19}$$

We can therefore write

$$\Delta S_{\text{repair}} = -f(t)S_{\text{damage}} \tag{C.20}$$

and treat $f(t)$ as a constant over the repair time period when we look at the entropy repair rate.

Equation (C.18) can be compared to the well-known RC circuit shown in Figure C.3. The notion that the body charges up (switch B) in order to energize the repair area, before discharging energy (switch A), has a similar differential equation given by

$$I = \frac{dQ}{dt} = -\frac{Q_o}{RC}\exp\left(-\frac{t}{RC}\right). \tag{C.21}$$

Comparing this model to Equation (C.18), $S_{\text{repair}} = -S_{\text{damage}}$ and charge flow equates to entropy flow. The fractional repair and repair time are $f(t) \sim 1/R(t)C(t)$ and I is then the entropy current. For $f(t)$ to decrease, R and C would need to increase. We would need at this point a biological model for R and C. Perhaps it makes sense that the resistance of the internal body increases with time as the unrepaired entropy disorder builds up over a person's lifecycle. After all, we associate entropy production with resistance. C would also increase if the dielectric constant increases; this needs some thought on why that would occur if disorder were to increase over time. It may be that only R increases and C is constant.

References

[1] Schrödinger, E. (1944) *What Is Life?* Cambridge University Press, Cambridge.

[2] Schnohr, P., O'Keefe, J.H., Marott, J.L., Lange, P. and Jensen, G.B. (2015) Dose of jogging and long-term mortality. *Journal of the American College of Cardiology*, **65** (5), 411–419.

[3] Becker, R.O. and Selden, G. (1985) *The Body Electric, Electromagnetism and the Foundation of Life*, William Morrow Publisher, New York.

Overview of New Terms, Equations, and Concepts

Key Words

Degradation, aging, thermodynamics, physics of failure, accelerated testing, reliability, thermodynamic work, entropy, entropy damage, free energy, fatigue, conjugate work, fatigue damage spectrum, noise analysis, noise damage, white noise, corrosion, cyclic work, Miner's Rule, heat engine damage, negative entropy, spontaneous negative entropy, human aging, environmental profiling, transistor aging, thermal cycle, shock and vibration, log time aging, complex systems degradation, acceleration factors, wear, creep, diffusion, aging law influence on reliability distributions, CAST goals.

New Terms, Equations, and Concepts

This book offers a lot of new work. We provide here a list for the reader.

Unique Terms

Entropy damage, entropy damage threshold, spontaneous negative entropy, repair entropy, cumulative accelerated stress test (CAST) equations and goals

Unique Equations

2.2, 2.10, 2.14, 2.23, 2.24, 2.36, 2.42, 2.124, 2.128, 3.21, 3.25, 3.31, 4.15, 4.16, 4.19, 4.22, 4.23, 4.26, 4.29, 4.30, 4.33, 4.47, 5.3, 5.4, 5.6, 5.7, 5.10, 5.11, 6.7, 6.8, 6.15, 7.7, 7.8, 7.15,

---

Thermodynamic Degradation Science: Physics of Failure, Accelerated Testing, Fatigue, and Reliability Applications, First Edition. Alec Feinberg.

7.16, 7.20, 7.24, 7.25, 7.31, 7.36, 7.37, 8.17, 8.18, 8.21, 8.22, 9.15, 9.26, C.7–C.10, C.16–C.20.

Please see also Table 1.5, Table 4.3, Miner's rule derivations, and acceleration factor equations.

Unique Concepts

Second law phrasings in terms of degradation, four main categories of aging, thermodynamic aging state, thermodynamic reliability, entropy noise principle, system entropy random process postulate, mesoscopic noise entropy measurements, system undergoing parametric degradation will have its noise level increased, environmental noise, physical implication of log time aging connected to the lognormal distribution (Equation (9.8)), physical implication of power law aging connected to Weibull β (Equation (9.26)), environmental profiling, CAST (cumulative accelerated stress test) goals and equations, human engine, Figure C.1, RC repair model.

Please Do Cite the Author

The list provided here of new terms, equations, and concepts are viewed by the author as original; the author kindly requests a reference when interested readers are referring to these in their future publications.

Index

accelerated test, 1–8, 204–205, 207–209, 212–214, 216–223
  CAST goals, 222–223
  conferences, 207
  environmental profiling, 222–223
  high temperature, Arrhenius model, 125, 209–212, 221–223
  multi-test plan, 1–8, 204–205, 209, 220–223
  qualification testing, 2–8, 209, 221–223
  salt fog, 208
  temperature cycle, 213, 216 217, 221 223
  temperature–humidity–bias, 214–216, 221–223
  vibration, 213, 217–220
acceleration factor, 87–88, 95–96, 98, 100, 102–103, 125, 126, 128–130, 161, 209–217, 221–223
  creep, 98, 212
  damage, 88
  diffusion, 161
  Peukert's law batteries, 126
  temperature, Arrhenius, 125, 209–212, 221–223
  temperature-humidity, corrosion, Peck, 128–130, 214–216, 221–223
  temperature-humidity, local RH, 215
  thermal cycle fatigue, Coffin-Manson, Norris-Lanzberg, 102, 213, 216–217, 221–223
  time compression, definition, 88
  vibration fatigue, 103, 213, 217–220
  wear, 100, 213
activation, 2, 125, 129, 135, 140, 211–212
  Ea, 125, 129, 140, 211–212
  energy, Ea, table of values, 140
adiabatic, 9, 82
aging, categories, 2
aging ratio, 31–32, 37
Allan variance, 37
angular momentum, 8
angular velocity, 8
Archard's wear equation, 99–100, 144–146
Arrhenius, 33, 125, 134, 136, 140–141, 209–212, 221–223
  barrier probability, 134
  law, 125, 209–212, 221–223
  TAT model
    large parametric change, 140–141
    small parametric change, 136
availability, reliability definition, 184–185, 189, 209

Thermodynamic Degradation Science: Physics of Failure, Accelerated Testing, Fatigue, and Reliability Applications, First Edition. Alec Feinberg.

availability, useful work, 56
average failure rate $\gamma(t)$ave, definition, 185

Basquin's equation, 104
bathtub curve, 163, 167–168, 173–174, 187–188
  creep rate, three stages, primary, secondary, tertiary, 173–174
  Power law model of, 167–168
  reliability distribution fit curve, 187–188
  Weibull beta, 168
batteries, 119–120, 124, 126–128
  Nernst Equation, 127–128
  primary batteries, 124, 126–128
  secondary batteries, 119–120
    damage, 120
    DOD, 119–120
Boltzmann constant, 34, 125
Boltzmann distribution, 138

capacitor leakage model, 149–150
CAST (cumulative stress test) goal, 99, 101–102, 104, 107–108, 222–223
  creep, 99, 108
  fatigue, Miner's rule, 108
  thermal, 108
  thermal, corrosion, 222–223
  thermal cycle fatigue, 102, 108, 222–223
  vibration, 104, 108
  wear, 101, 108
catastrophic failure, 16
censored Data, 199
central limit theorem, 160, 197
charge, 8, 53–54
chemical differential rate law, 129
chemical potential, 8, 9, 121
chi-squared confidence Test, 204–206, 212
clausius inequality, 15
Coffin-Manson AF model, 102, 213, 216–217
compressibility, 9
confidence interval for normal parameters, 195–197
confidence testing, 203–206
controlled mass, 4
corrosion, 118–130
  acceleration factor, Arrhenius, 122, 125
  chemical process, 121
  current, backward, forward, net, 121–123, 130
  damage, 122, 126
  Faraday's Law, 122
  Nernst equation, 128
  Peukert's law, 126
  rate, examples, different metals, 123–125
  rate in microelectronics, 128–130
  secondary batteries, 119–120
  thermodynamic work, 119
  uniform corrosion, 127
Cpk index
  and its relation to yield, 197–198
creep, 96–99, 172–173, 212, 222–223
  acceleration factor, 98
  accumulated damage, 98–99
  CAST goal, 99, 222–223
  rate, aging power law, 96, 173, 212
  rate, steel, 96
  similar to bathtub curve, 173
  three phases, 172–173
  time to failure, 98
cumulative distribution function (CDF) F(t) to fit data method, 184, 199
cumulative failure rate $\lambda(t)$cum definition, 184
  see also specific CDF
cumulative strength, 26, 27
current, 8

damage, 1, 2, 25–27, 79, 85–88, 94–95, 120–122, 126 *see also* Miner's rule, cumulative damage; creep, wear, vibration fatigue, thermal fatigue
  acceleration factor, 88
  corrosion, 122, 126
  cumulative, 26, 27, 79
  cyclic damage, 86–87, 94–95
  damage, 26, 27, 79, 86–88, 94–95
  effective, actual, 94–95
  measurable damage, 86
  non-cyclic damage, 87
  ratio, 88
  secondary batteries, 120–122
  thermodynamic, 1, 2
  work damage concept, 85–87
damping loss factor, 107
density, 9
differential entropy failure rate, 50
diffusion, 2, 157–164
  acceleration factor, 161
  concentration, 159
  entropy damage, 161–162
  equation, 160, 163
  equilibrium conditions, 158–159

diffusion (*cont'd*)
  Gaussian probability, 159–160
  maximizing entropy, 158
  package moisture diffusion, 162–163
  random walk, 159–160
  spontaneous, 157–158
  variance, 160
disorder, 1, 81 *see also* entropy and disorder
disordered energy, 38
displacement, generalized, 7

efficiency, human, 231
efficiency, inefficiency, 57, 83–85
elastic stress limit, 12
elasticity (Young's modulus), 9, 12, 168
electric field, 7–9, 140, 151, 157, 163
electrolyte, 126, 128–129, 148, 157
engine, 7, 18, 80–85, 225–229
  Carnot cycle, 81
  Carnot cycle efficiency, 84
  damage, efficiency loss, 84–85
  efficiency, 83–85
  heat engine, 81–83
  human, 18, 225–229 *see also* negative entropy
  MPG, efficiency, 85
enthalpy, 9, 60
entropy, 3, 4, 6, 10–18, 23–25, 28–29, 31–32, 34–35, 37–40, 50–55, 65, 77, 81, 181, 224–225, 231–233
  change (ds or Δs), 4, 10, 11, 14, 77
  continuous and discrete, 35
  damage, 12, 13, 17, 23–25, 31–32, 77, 224–225, 231–233
  damage axiom, 11, 13
  damage resolution, 24
  damage threshold, 12
  damage, cumulative, 23, 25
  damage, system level, 29, 31–32
  differential, 35
  δQ/T, 15, 18, 51, 65
  environmental entropy damage, 81
  exchanged, 11
  failure rate, differential entropy, 50
  flow, 11
  generated, 28
  maximize principle, 51–55
  mesoscopic noise, 34
  negative entropy, 17, 18, 224–225, 231–233
  noise autocorrelation function, 39
  noise principle, 33
  non-damage, 4, 11, 24
  of the Universe, 6, 11
  production, 55
  random process postulate, 37–38
  randomness mesoscopic measurements, 40
  repair, 17, 18, 224–233
  statistical definition, 34–35, 181
  statistical definition with microstate model, 181
environment damage, 6, 11, 81
environment, definition, 4 *see also* most of Chapter 1 and 2 for usages
equilibrium, 5, 16, 52–55, 158–159
  diffusion equilibrium, 55, 158–159
  equilibrium thermodynamics, 5, 16
  equilibrium with Charge Exchange, 53
  thermal equilibrium, 52–53
exponential distribution
  R(t), CDF, PDF, failure rate, uses of, 188–189
extensive variables, 7, 8, 20

failure rate, 24, 50, 137, 167–168, 171–174, 184, 186, 187
  conversions, 186, 187
  for series system, 186
  λ(t), definition, 184 *see also* specific distributions
  time dependence and independence, 186
Faraday's law, 122, 127
fatigue damage spectrum (FDS), 108–111
  random vibration, 110–111
  sine vibration, 109–110
fatigue limit, 10, 11
first law of thermodynamics, 6, 51, 55, 82, 83
  for cyclic process, 83
force, conjugate, 7
forced processes, 2
Fowler–Nordheim tunneling leakage/breakdown, 151
free energy, 5, 13, 33, 134
  barrier, 33, 134
  relative minimum, 134

gas constant, R, 125
Gaussian, 35–36, 169
  differential entropy, 35–36
  pdf, 35, 169, 192–193 *see also* normal distribution
  white noise, 35–36

Generalized Power Law Acceleration Factors, 214
Gibbs free energy, 8–9, 13, 57, 60–62, 65, 67, 78, 118–119, 121, 127
  bounds useful work, non-compressive, 61, 118
  damage, 121
  full expression for work, 67
  non-mechanical work, 62
growth, self-repair, 229–231

hardening, strain, 96
hardness, 9, 99
heat, 6, 9–11, 13–15, 17, 18, 50, 51, 58, 61, 63, 65–66, 82–85, 118, 144, 207
  capacity, 9
  path δQ, 15
  reservoir, 65–66, 82
Helmholtz free energy, 9, 13, 18, 41–42, 57–59, 62–63, 65–67, 78, 138–139
  maximum useful work, reversible work, 59
  measurable using work, 78, 86
  mechanical work, 58
  opposite of entropy concept, 13, 66
  partition function relationship, 138–139
  reversible work, 78
human engine, 18, 225–229 *see also* negative entropy
human heart
  noise, failure, S-N curve, 41–42, 228

inflection point, 202
intensive variables, 7
internal energy, 6, 7, 13, 30, 58, 62, 67
internal irreversibility, 10
irreversible, 2–4, 5, 9, 11, 15, 28, 52–53, 59, 62, 65
isentropic, 9
isobaric, 9
isochoric, 9
isolated system, 10
isothermal, 6, 9, 82

kinetic frictional model, 145

Legendre transform, 58, 61
logarithmic-in-time aging, 135–138, 146–147, 150–154, 162, 168–171
  bipolar transistor beta aging, 150, 152
  connection to log-normal distribution, 168–171
  creep model, 147
  diffusion, 162
  field-effect transistor, 154
  graphical shape, 137
  power law resemblance, 137–138, 169
  wear model, 146
lognormal distribution, 168–171, 175–178, 194–195, 199–201
  Arrhenius life-stress model, 175–177
  CDF, error function, 170
  connection to log-time aging, 168–171
  fitting catastrophic data example, 199–201
  graphical sigma, sigma time dependence, 171, 178
  mean, 170
  PDF, 170
  R(t), CDF, PDF, failure rate, 194–195
  vibration life-stress model, 175–177

Maclaurin series, 136, 147
macroscopic system, 3, 34
magnetic field, 9
magnetic flux, 8
magnetic intensity, 8
mass, 4, 7–9, 11, 23, 122–123, 128–129, 144–146, 157
mean time to failure (MTTF) or mean time between failures (MTBF), 17, 185, 187, 189, 199–200, 204
mesoscopic, 27, 33–34, 37, 40
  measurement, 34–37
  noise entropy, 34
  system, 34
microstates mode, 181
Mile's equation, 106, 110
Miner's rule, 25–27, 87, 94–96, 101, 108–110, 119–120, 227
  acceleration factor, 95–96, 101–102
  corrosion, 121
  derivations, cumulative and cyclic damage, 25–27, 87, 94–95, 227
  example, 26, 93–95
  FDS, 108–110
  non-cyclic damage, 27
  secondary batteries, 119–120
mixed modal analysis
  subpopulation, main population, 201–203
momentum, 8

negative entropy, 17, 18, 224–233
  perfect human engine, 225–229
  spontaneous negative entropy, 224–225, 231–233

Noise Analysis, entropy, 33–51
Allan variance, 37
autocorrelation noise measurement, 48–49
autocorrelation, function, 40–42
delta function, 43
disordered energy, 38
environmental noise, pollution, 50
exponential, 43
Flicker (pink) 1/f, brown $1/f^2$, noise, 45–48
Fourier transform, 42
Gaussian, 43
Gaussian white noise, 35–36
human heart congestive heart failure, 41–42
Johnson–Nyquist noise, 49–50
measurement system, 48
mesoscopic noise, 34
mesoscopic noise entropy, 34
noise analysis, 36
noise entropy autocorrelation function, 39
noise principle, 33
Nyquist theorem, 49–50
PSD (Power Spectral Density), 42
random process postulate, 37
stationary, non-stationary, 40
system noise degradation, 33–35
vibration, 35–37
white noise, 43, 45
Wiener–Khintchine theorem, 43
non-equilibrium thermodynamics, 4, 5, 16
normal distribution, 191–194 *see also* Gaussian distribution
Norris–Lanzberg AF model, 102

operating system, definition, 21, 33
order, 181–182
organization and entropy, 181–182 *see also* theory of organization

parametric aging state, 14, 136, 140
parametric failure, 14, 16, 27, 140, 169–175
partition function, 138–139
path dependent, 6, 15, 30, 63–65, 78, 82, 85–86
Peck's humidity AF, 129–130
Peukert's law, 126
physics of failure, 1–3, 79, 173, 184, 209
polarization, 8
Poole–Frenkel, dielectric leakage/breakdown, 151
power law aging models, 163, 212 *see also* creep, thermal cycle, fatigue, Peukert's law
power loss, 65, 135, 169
power spectral density (PSD), 36, 42–48, 104–106, 110–111, 213, 219–220
power, dW/dt, 8, 16, 34, 65, 135, 169
pressure, 8, 56
probability density function (PDF) f(t), definition, 35, 184 *see also* specific PDFs for Weibull, lognormal, exponential, normal

Q, 49, 105–107, 110
quasistatic process and measurement process, f, 5, 9, 28–30

redundancy, 185
refrigerator, 80, 84
reliability function R(t), definition, 184 *see also* specific CDF
reliability software to aid the reader, 183–184
repair entropy, 17, 18, 224–233
repairable-reversible, 6
resistor aging, 30
Resonance, 49, 105–107, 110, 208, 218, 220
Reversible process, 4–5, 9–11, 14–16

S-N curves
and equation, 10, 104, 119–120, 213
human heart compared to metal fatigue, 228
scanning electron microscope (SEM), 34
Schottky effect leakage/breakdown, 151
second law, 2, 6, 9–11, 13, 17–18, 51–61, 63, 67, 80, 84, 158, 224–225, 229–232
combined with first law, 51, 54–61, 63, 67, 158
damage, $\Delta S_{damage} \geq 0$, 10
in entropy damage terms, 2, 10, 11
in free energy terms, 13
self-repair, growth, 229–231
specific heat, 9, 30–32
spontaneous process, 3, 5, 17, 66, 157, 159
spontaneous repair, 6, 231–233 *see also* negative entropy
state variables, 7–8, 29, 40–41, 50, 77
stokes integral theorem, 80, 94
strain, 8, 12, 79, 86–87, 96–97, 101, 103
plastic, 79, 86–87, 96, 101, 103
stress, 8, 12, 96–97, 101
creep, 96, 172–173
thermal cycle fatigue, 101–102, 122–123, 213, 216–217

system
  definition, 1, 4 *see also* Chapters 1 and 2 for common usage System
system vibration damage, example, 36–37

TAT model, 135–138, 140–141, 144–147, 152, 154, 171
  activation wear model, 144–146
  bipolar transistors, 152
  creep model, 147
  field-effect transistor, 154
  graceful degradation, 135
  influence on lognormal distribution, 171
  large parametric change, 140–141
  logarithmic-in-time aging, 135–138
  small parametric change, 135–138
  thermally activated time dependent model, 135–138
temperature, 7, 9, 29–33, 49–50, 53, 55, 56, 59–64, 66, 82–85, 96, 101–102, 118–119, 125–131, 134–141, 145–148, 151–152, 154, 158, 161–163, 175–176, 209–211, 213, 215–217
tension, 8
theory of organization, 181–182
  influence on manufacturing, reliability and quality, 181–182
  probability of success, 181–182
  reducing complexity, minimize variability, 181–182
  stability, for materials, 181–182
thermal cycle fatigue, 101–102, 122–123, 213, 216–217
  acceleration factor, 102, 213, 216–217
  CAST goal, 102, 222–223
  Coffin-Manson AF model, 102
  damage, 101–102
  Norris–Lanzberg AF model, 102
thermally activated, 33, 124, 134–135, 139–140, 144, 150–151, 162–163, 168, 210
thermo-chemical E-model leakage/ breakdown, 151
thermodynamic aging states, 14, 29
thermodynamic degradation science, 1
thermodynamic Potentials, 6, 7, 9, 13, 55–67, 78
  available work, 55–56, 64
  chemical potential, 57, 61–62, 158
  enthalpy, 60, 62, 67
  example, 62
  Gibbs free energy, 9, 13, 60–62, 67
  Helmholtz free energy, 9, 13, 58–59, 62–67, 78
  internal energy, 6, 7, 13, 58, 62, 67
thermodynamic state variables, 7–8, 29, 38, 40–41, 50, 77
transistor aging, 148–154
  beta degradation, 151–152
  bipolar case, HBT, 148–152
  capacitor leakage model, bulk and surface, 149–150
  field-effect transistor, MESFET, 152–154
  gate leakage, base leakage, drain-source resistance, 151–153
  transconductance degradation, 153
transmissibility, 105

Variance, Allan, true, 37
vibration, 103–107, 222–223
  CAST goal, 104, 222–223
  cyclic fatigue work, 103
  fatigue acceleration factor, 103, 104
  fatigue damage, 102, 103, 105, 107
  fatigue failure at resonance, 105–106
  G, Grms, 103
  PSD, power spectral density, 104
  random vibration, 104–107
  resonance, Q, transmissibility, 105
  sine vibration, 103–107
voltage, 7–9, 27, 33, 46–47, 49, 119–121, 126, 135, 137, 140, 149, 150, 153
volume, 7–9, 13, 30, 32, 48, 52–53, 58–59, 65, 66, 81–82, 85, 118, 160

wear, 99–101, 144–146, 222–223
  abrasive, 99
  acceleration factor, 100
  activation log-time wear model, 144–146
  Archard's coefficient, 99
  Archard's equation, 99–100, 144–146
  CAST goal, 101, 222–223
  damage, 100
  time to failure, 100
Weibull distribution, 167–168, 171–175, 199–201
  beta, meaning of, bathtub curve, 167–168
  connection to power law aging, 171–175
  creep parameters and beta, 174
  failure rate, 167
  fitting catastrophic data example, 199–201
  power law concept, 167
  R(t), CDF, PDF, failure rate, 190–191

work, thermodynamic, 6–8, 57, 63–65, 67, 78, 81–88, 98–99, 101, 103–104, 119, 121, 125, 151, 158–159
  actual work, 57, 64–65, 78, 85
  available Work, 55–56, 64
  chemical work, 119, 121, 158
  conjugate work
    creep, 98
    diffusion, 159
    electrical, 119, 121, 125, 151
    variables, 8, 78
    vibration, 104
    wear, 99
  creep, 98
  cumulative work damage, 86–88
  cyclic work, 80
  cyclic work for secondary batteries, 119
  engine, 81–84
  irreversible work, 57, 64–65, 78, 85
  path dependent, δW, 6, 63–65, 86
  power loss, 65
  reversible work, 56–57, 64–65, 78
  thermal cycle fatigue, 101
  thermodynamic work, 6, 7, 67
  useful work, 57, 64–65
  vibration fatigue, 103
  work efficiency, inefficiency, 57, 64–65

yield point, 12
Young's modulus, 9, 12, 168